# The American Trap

My battle to expose America's secret
economic war against the rest of the world

Frédéric Pierucci with Matthieu Aron

Translated from the French by Deniz Gulan

HODDER

First published in Great Britain in 2019 by Hodder & Stoughton
An Hachette UK company

This paperback edition published in 2020

1

A CIP catalogue record for this title is available from the British Library

Paperback ISBN 9781529326871
Hardback ISBN 9781529326864
Trade Paperback ISBN 9781529353617

Typeset in Bembo by Hewer Text UK Ltd, Edinburgh
Printed and bound in Great Britain by Clays Ltd, Elcograf S.p.A.

Hodder & Stoughton policy is to use papers that are natural, renewable
and recyclable products and made from wood grown in sustainable
forests. The logging and manufacturing processes are expected to
conform to the environmental regulations of the country of origin.

Hodder & Stoughton Ltd
Carmelite House
50 Victoria Embankment
London EC4Y 0DZ

www.hodder.co.uk

*Dedicated to my wife and children*

# Contents

Appendices

*Out of respect for their privacy, the names of the members of Frédéric Pierucci's family, as well as those of his relatives and close friends, have been changed.*

# Author's note

Alstom, as it once was, has ceased to exist. I thus wish to pay tribute to my former colleagues, workers, engineers, technicians, sales staff and project managers, who strived for decades to produce high-performance state-of-the-art products that were the envy of our rivals and helped guarantee France's energy independence.

Let there be no misunderstanding. Though this book highlights a series of disastrous decisions, I still cherish the memories of the solidarity and team effort that shaped my daily existence for twenty-two years.

# Foreword

This book tells the story of what I discovered – through bitter experience – about how the United States of America, in the guise of the global fight against corruption and terrorism, conducts its underground economic warfare.

Over a period engulfing more than ten years, the United States has succeeded in destabilizing the largest European multinationals. The American Department of Justice (DOJ) targets their senior managers, incarcerates them if necessary and nets billions of dollars in fines by coercing those companies to plead guilty.

Since 2008, thirty corporations have each paid out fines above $100 million to the American Treasury. Sixteen of them are European (five are French, three are German, two are British) and just seven are American.

European corporations will have shelled out a total of more than $6 billion, with American corporations paying three times less. French corporations alone have already paid a total of nearly $2 billion and six of their executives have been indicted by the American judiciary.

I am one of them.

I refuse to remain silent any longer.

This is my story, from my perspective, as I lived and experienced it.

# I

# Shock and disbelief

I have become an animal. Draped in an orange jumpsuit, chains are wrapped around my chest, handcuffs lock my feet and hands. I can hardly walk or breathe. I am a tied-down brute. A creature snared in a trap.

Last night they threw me into a cell where the stench is so strong, I feel giddy. No windows, just a tiny slit beyond which I guess is a courtyard. Then there is the noise, the quarrelling, the fights, the screams, the incessant screams. It's a nightmare. I'm hungry and thirsty, very thirsty. Water hasn't touched my lips for eight hours, since the announcement came, turning my life upside down.

A seemingly insignificant polite message by a Cathay Pacific flight attendant with a perfect English accent. Her dulcet tones forebode disaster: 'Mr Pierucci is kindly requested to present himself to the cabin crew upon landing.'

The Boeing 777 I was on had just touched down at New York's JFK airport. I had left Singapore at dawn. After a stopover in Hong Kong on a journey of more than twenty-four hours, I was deadbeat. It was 14 April 2013 and 8 p.m. The pilot had completed his flight plan to perfection. The aircraft was just arriving home when the voice resounded out.

Should I have been suspicious? Right now, I'm knocked out by jet lag, even though I'm used to long-haul flights. At forty-five years old, after having been based in Algiers, Manchester, Hong Kong, Beijing, Windsor (Connecticut), Paris and Zurich, I currently live in Singapore. During my two decades of globetrotting for

Alstom, I am used to receiving this type of message, whether it is to report a rescheduled meeting or a mobile phone forgotten during a stopover.

So without any particular misgivings, I presented myself to the chief flight attendant. The young woman looked a bit embarrassed. The aircraft door opened and, with a shy and fleeting gesture, she points out a group of people who are waiting for me when I get off the plane.

A woman, two or three men in uniform and two in plain clothes. The woman politely asks me to confirm my identity, then orders me to get off the plane. I barely have time to utter my name before one of the uniformed men grabs my arm, wrenches it behind my lower back, then takes my other arm and pushes it up onto my back, handcuffing me: 'Frédéric Pierucci, you're under arrest.'

Totally dazed, I don't react. I simply do not resist. Later on, I often wondered what would have happened if I hadn't left the plane? If I had refused to disembark? Would they have been able to arrest me so easily if I had not yet set foot on American soil?

I complied without flinching. Unwittingly, I made it easier for them, as we were still in theory in the international zone, even on the walkway after leaving the aircraft.

Right now, I'm standing here in handcuffs. Shocked and astonished, I demand an explanation. The two plain clothes agents tell me they're FBI detectives.

'We've only had one instruction, to arrest you as you leave the aircraft and escort you to FBI headquarters in Manhattan. When you arrive, a district attorney will brief you.'

I have no choice but to settle for these meagre words and follow them through the airport, my hand shackled behind my back like a gangster with the two marshals in uniform watching over me like hawks.

I am conscious of the heavy gaze of the other passengers on me. After a few yards, I realize that, to keep my balance, I can only take tiny steps forward, which is no mean feat for someone of six foot one weighing 200 pounds. I feel and look like a freak. It could be a movie scene in which I'm playing the role of D.S.K.,* who only two years earlier was just like me, on the streets of New York, hand-cuffed and escorted by the FBI.

Right now, though, I'm more astonished by what is happening to me than concerned. It is obviously a mistake or a misunder-standing. The FBI has mistaken me for someone else and, after a few checks, my name will be cleared and life will be back to normal. There has been an increase in these types of incidents at New York-JFK in recent years.

My guards lead me directly into a small room. I know this place by heart. It is where the American authorities conduct their metic-ulous inspection of passports of aliens potentially posing a risk. During the second Iraq war in 2003, due to France's refusal to fight alongside the US forces, French businessmen have been held in this airport annexe for hours, waiting for US officials to finally agree to let them enter US territory.

Today, everything proceeds with the speed of light. After check-ing my papers, the two inspectors whisk me out of the airport and into an unmarked car. I then realize that I am indeed the person they were waiting for. This is not one of those absurd stories where a member of the public is mistaken for a potential terrorist or a hitman. But why me? What do they want from me? And what have I done?

I don't need to spend long sifting through my life. I have noth-ing to reproach myself for in my personal affairs. That leaves my job at Alstom. But the idea that this brutal arrest could be related to my professional activities seems equally improbable to me. At full

---

* Dominique Strauss-Kahn, French politician, former head of the IMF from 2007 to 2011, who was arrested on allegations of the sexual assault of a maid in the New York Sofitel hotel in 2011.

speed, I perform a quick checklist of current files in my head. None of the recent projects I have dealt with since being in Singapore, where I took up my duties ten months earlier as global president of the boiler division, seem to me to be tainted by the slightest suspicion.

However, I do know that Alstom is regularly under investigation for corruption and that an investigation was opened by the American Department of Justice several years ago. The group of Alstom companies was accused of allegedly having paid bribes to secure numerous contracts, including one for a power plant in Indonesia. At the time, I had been working on this project and Alstom had indeed used external 'consultants' to secure the contract. But these facts date back to 2003 and 2004, and the Indonesian contract was signed in 2005. The negotiations took place almost ten years ago. In other words: old news. And to top it all, I was cleared by an internal audit, the type they conduct in similar cases. It was early 2010 or 2011, I don't remember the exact date, but what I am sure of, as the car races towards Manhattan and I become more and more nervous, is that two lawyers hired by Alstom interrogated me for about an hour. And to my knowledge, they were satisfied that I had complied with all the group's procedures.

Then, in 2012, I obtained a big promotion, giving me the position I currently hold as head of the boiler division, in charge of 4,000 people worldwide, for a turnover of $1.6 billion. In addition, Patrick Kron, CEO of Alstom, who since 2011 had sought to create a joint venture with the Chinese Shanghai Electric Company by merging the boiler activities of both groups, chose me to lead this future partnership at global level, based in Singapore.

Global president! Look at me now, locked in an FBI car, with sore aching arms. The metal of the handcuffs is cutting into my wrists. How can it be possible that the Indonesian case of 2003 to

2005, in which I only played a secondary role, is the reason for such brutal treatment? I am not Al Capone or even a small-time gangster! While I go over in my mind all the possible scenarios that could have led to my arrest, the investigators stop the car on the side of the road.

Ron and Ross, the FBI officers whose names I later learn, think I am 'nice, a stroke of luck'.

'Mr Pierucci, you're quiet, you're not yelling, you're not struggling. Besides, you're polite. It's pretty rare that we land on someone like you. We're gonna give you a break.'

They take off my handcuffs, bring my arms – which had long since seized up – back onto my lap, and handcuff my hands in front of me. The relief was immense.

Tonight, the traffic is flowing smoothly and we arrive in Manhattan in less than forty minutes, right in front of the FBI headquarters, where the car plunges into the basement.

We approach an elevator. The police officers prompt me to enter it, 'backwards'. I look at them in amazement. I worked in the United States for seven years, from 1999 to 2006, and consider my English to be perfect. But right now I'm in doubt: do I really have to get in that elevator backwards?

'It's for security reasons Mr Pierucci,' Ron explains to me. 'You have no right to see which button we press. At FBI headquarters, you will not know which floor we're taking you to, or in which office you're going to be questioned.'

So, I am being led to a mystery floor. After passing through several armoured doors, we end up in a modest-looking office. A room containing no furniture, save a table and three chairs in the middle, and embedded in the wall, a long iron bar to which my guards handcuff me. They leave me alone for a few moments. Suddenly the door opens, and a new investigator appears.

'Hello Mr Pierucci. My name is Seth Blum. I'm investigating corruption cases within the FBI, targeting Alstom, notably the

Tarahan case in Indonesia. I cannot tell you anything more because, in a few minutes, the prosecuting attorney in charge of the investigations will be coming to question you himself.'

Without further ado, Seth Blum quietly and politely leaves the room.

# 2

# The US attorney

So that's what it was all about. In the car, I didn't want to believe it. Too long ago, too remote, too disconnected from present-day challenges. But I must face facts. This goddamn Indonesian case, the Tarahan power plant located on the island of Sumatra, has caught up with me. When the tender to construct it was launched in 2003, I was not yet based in Singapore. I was operating from the United States as global sales and marketing director of one of Alstom Power's divisions.

To compete, we formed a consortium with the Japanese company Marubeni. At that time, Alstom was undergoing severe financial difficulties, and on the verge of bankruptcy. Although this contract appeared relatively modest ($118 million, including some $60 million for Alstom), token importance was attached to it. The contract for the power plant on the Indonesian island of Sumatra, just a tiny dot on the world map, would breathe new life into Alstom and restore the company's reputation.

In the FBI interrogation room – Seth Blum having just told me the reasons for my arrest, and with me waiting for the attorney to explain more – I rewind to 2003 and remember how hard we had to battle to win Tarahan. We should not overlook the fact that in some countries around the world at that time, bribes were a common if not accepted practice. I also know that Alstom used two intermediaries, referred to as 'consultants'. I certainly did not initiate their recruitment, but I knew of their existence.

I barely have the time to revisit that era when the door opens and a man enters, accompanied by Seth Blum. He is around thirty-five years of age. Short, curt, in a manner I saw as arrogant, he spews out his speech with the flow of a machine gun.

'Mr Pierucci, my name is David Novick and I am assistant US attorney for the district of Connecticut in charge of the case against Alstom, your employer. You have been arrested in connection with the corruption proceedings we have instituted against your company. I am leading this investigation. You are accused of co-conspiring in acts of corruption in favour of an Indonesian government official in connection with the Tarahan contract. This constitutes corruption of a foreign public official and violates the Foreign Corrupt Practices Act.* We have been investigating Alstom's worldwide practices for three years now. As required by US law, we immediately informed your group of this. And yet, despite its promises, your company has persistently refused, since 2010, to fully cooperate with the Department of Justice.† Alstom hasn't respected any of its commitments. Not one!'

He seems furious. I am tempted to reply that I am neither the CEO of Alstom, nor its general counsel. I may well be a senior executive, but I am not a member of the Board or of the executive committee.

'I . . .' But the attorney interrupts my thought process.

'Mr Pierucci, I strongly advise you not to call your company. We'd like you to do things for us . . .'

At that moment, my thoughts start to get muddled. What exactly is this prosecuting attorney asking me?

'We'd like you to help us against Alstom and its management. We are well aware of the position you hold at Alstom today and the

---

* FCPA: a United States Federal Law of 1977, to combat the bribery of public officials abroad.
† DOJ: United States Department of Justice.

one you held at the time of the Tarahan events. We know very well that you were not one of the decision makers in the Indonesian case, but you nevertheless knew exactly what was going on. We want to prosecute Alstom's top management and, notably, its CEO, Mr Kron. We are therefore asking you not to inform them of your arrest. That's why you should not call them and so, for the time being, you should give up the help of a lawyer, but of course it is your choice. Do you understand?'

Well, no, I don't understand. Or rather, yes, I get the sense of the deal that is looming. David E. Novick is suggesting that I become his informant within the firm. I am jet-lagged, and have not slept for twenty-four hours, and I am still at the mercy of the handcuffs that shackle me to this iron bar. What should I comprehend exactly? He makes no attempt to help me, remains vague, makes no precise request except to repeat incessantly that I must not tell anyone. To me, this is just unthinkable.

As he tries to silence me, I recall being in one of Alstom's training sessions for senior managers. Ironically, this was shortly before my arrest. The seminar focused on the legal risks of our profession. We were given a small business-type card with the phone numbers to call in the event of arrest, including Keith Carr's, the group's general counsel. We were told to carry this little card with us at all times, and should we ever be confronted by a judge or a police officer, never to waver from the two rules instilled during these coaching sessions:

1. Do not say anything.
2. Call Alstom's legal department, which will immediately send a lawyer to the employee in question.

I remembered the lesson and resolved not to fall into the attorney's trap; at least that's what I believed at that moment. In utter loyalty to Alstom and without imagining for a second what it would cost me, I applied the precepts given by my company's legal department and declined. I needed to alert them, and that's what I explained to the attorney.

'Listen, I have never been arrested before, I don't understand what you want. I therefore ask you to allow me to inform my company as well as my consulate.'

Poker faced, the attorney signals to an investigator, who hands me my BlackBerry, confiscated during my arrest. I immediately attempt to reach Keith Carr. In Paris it is 5 a.m., and the phone just keeps ringing. However, I am able to reach Tim Curran, the head of Alstom's power division in the United States, with whom I had an appointment the next day in Windsor, Connecticut. I briefly summarize the situation to him. He is flabbergasted.

'What is happening to you is unbelievable. This is completely absurd. We'll get you out of there fast. I'll notify headquarters immediately.'

Tim Curran comforted me somewhat. The attorney has gone and two police officers search and conduct a complete inventory of my suitcase. I still have the right to make one more phone call. I hesitate whether to call Clara, my wife, and then I decide not to. Why worry her? At this point, I am absolutely convinced: it is only a matter of a few hours. I'll soon be released. David E. Novick may well have shown himself to be uncompromising, but no matter how much he insists that Alstom has been under investigation for almost three years; that the group has not responded to any of the Department of Justice's injunctions; that it has not heeded any of the questions asked, and has played dead – I just don't believe it. Or rather, I do not want to believe it. I am firmly convinced, and would put my head on the line to prove it, that my company will come to my aid as soon as possible. I know I can count on the CEO; he has total faith in me.

A few weeks before I left for New York, I had dinner with Patrick Kron. He had invited me, plus a few Asian-based executives, to a sumptuous reception in Singapore, at a very special venue: the Marina Bay Sands, the most sought-after restaurant in the capital. Footage of this establishment has circulated worldwide. It is a rather wacky building with a gigantic flat roof on the fifty-seventh floor, which projects over the ocean like the bow of a boat. Keith Carr was also present. For several years now, Alstom had been developing a large part of its energy activities in Asia, to the extent that Kron was even considering partially transferring the group's headquarters to Singapore. An entire extra floor was rented at the end of 2012 to accommodate part of the Parisian workforce. Kron was a regular visitor to the premises and rumours started to circulate within the company. It was alleged that the CEO was considering taking up tax residence there. Admittedly, Singapore's tax rate is particularly attractive (maximum 20 per cent excluding allowances); the head of Alstom's Singapore office, Wouter Van Wersch, began prospecting the city in early 2013 to find a *pied-à-terre* for Patrick Kron.

To tell you the truth, I didn't really care about any of that. Though I wouldn't describe our relationship as being particularly close – even if we had got used to one another – Patrick Kron and I had what you might call a courteous relationship. In fact, one week before my trip to New York, I had accompanied him back to India to meet the managers of Reliance Industries, the largest private Indian conglomerate owned by the Ambani family. Patrick Kron is first and foremost a salesman, an outstanding negotiator, who quite happily globetrots alone, reaching out to his partners directly. He can be uncompromising, and sometimes almost rude, but he also knows how to ingratiate himself to clients. He made his mark among his troops in the field, rather than at headquarters in his own office, sometimes even bypassing the internal pecking order.

During that illustrious dinner at Marina Bay Sands, Keith Carr, loyal Kron fan, whom I have known for many years as he was previously in charge of the Power sector to which I reported, approached me and whispered:'Fred, do you remember the Tarahan case and the American investigation for which we also carried out our own investigation?'

'Yes, of course. Why, what's going on?'

'Not a lot; you have absolutely nothing to fear yourself. The internal investigation cleared you completely. But certain other individuals should be a little concerned.'

At the time, I thought it was a little strange that in the middle of a cocktail party he was talking to me about this case, whereas we had never mentioned it before, even in 2010 or 2011, when I was interrogated as part of the internal audit.

But right now, in the FBI office, this conversation springs to mind, presumably because I'm dialling Keith's number again.

At the second attempt, he finally picks up. The conversation is very short, but I remember every word.

'I don't understand, I don't understand . . . it's incomprehensible,' repeats Keith, who seems as stunned as I am. 'We are in the process of finalizing an agreement with the Department of Justice. It's just crazy what's happening to you.'

'Well, maybe, but the US attorney doesn't seem to know about this deal. Or he doesn't believe in it . . . He keeps telling me that I am here precisely because Alstom has failed to cooperate for three years and they have lost patience. Besides, you made it clear to me a few weeks ago that I had nothing to worry about. So why have they arrested me?'

'It's completely baffling. We're so close to an agreement that I shall be boarding a plane in a few hours. I am expected in Washington today, to strike a deal with the DOJ. Having said that, in light of what has just happened, I am reluctant to go to the United States. First, I will consult our lawyers. But don't worry. Above all, stay calm. Once I've contacted our law firm, we'll send someone to you. In the meantime, say nothing to the attorney, nor to the FBI. For tonight, it's a little late, but we'll get you out of there on bail tomorrow and then we'll see what strategy to deploy.'

And then he hung up. There is no doubt whatsoever: I'll hear from him in the early hours tomorrow; he won't desert me, he'll be at my side until the end. The company I have served for so many years simply would not abandon me. I'd have to be crazy or paranoid to even consider it.

As Keith's comforting words resonate in my ears, the prosecuting attorney returns to the interrogation room.

'You don't want to cooperate. Okay, it's your choice.'

'No, I'm willing to shed light on my role in this case, as I don't think I have much to reproach myself for, but I do need a defence lawyer as I have absolutely no knowledge of how the American justice system works, or indeed of my rights. I think any non-US citizen would react in the same way.'

My explanations make no impression on Attorney David E. Novick. Unwavering, he continues: 'So, I will transfer you to a prison in Manhattan. You will spend the night there. Tomorrow you will appear before a judge of the district court of Connecticut. You will have the right to meet with your counsel before this hearing. The judge will decide whether or not you remain in detention. If you wish, you can also make a phone call to your family to inform them.'

Stay calm. That's what Keith Carr advised.

Keep it together, play it cool. In any event, I have no other choice. Should I call Clara? The attorney seemingly prompts me to do so. Maybe he's trying to unhinge me? Of course, she'll be worried sick. And her distress will weaken my resistance. A classic when it comes to psychological pressure. The cops, I later learn, call it 'tenderizing the meat'. I think on my feet. I should be released tomorrow evening at the latest. The country that let out on bail Bernard Madoff or even O.J. Simpson for alleged homicide is not going to maintain me in detention, a French citizen, a corporate executive who, quoting the US attorney, did not have a 'decision-making role' in the Tarahan project targeted by the US judiciary. So no, I won't call my wife. I would rather tell her about my misfortune when I get out.

Politely, I decline David E. Novick's proposal. However, I demand that the French consulate in New York be notified. Novick immediately dials a pre-recorded number on his phone. Apparently, he has it all worked out. He knows exactly who to call at the consulate on a Sunday evening at midnight.

He passes me the handset and the person who answers is obviously 'on call'. The person I am speaking to asks my identity and specifies that he is just 'taking note'. Then Novick seizes the phone again to tell the consulate that I will be heard tomorrow, Monday, by a judge at the court in New Haven. The attorney has now finished with me for the evening.

Ron and Ross are back on the scene. They take an inventory of all my belongings (computer, laptop, clothing) in my small roller suitcase. Then we pass through the armoured doors, fingerprints are taken from all ten fingers, plus there is a photo session lasting a good half-hour. Then, back into the elevator, again entering it backwards, after which we drive to the nearby Manhattan prison.

The two inspectors don't leave my side during the whole admission procedure. Then, before leaving, Ron mutters: 'Good evening, Mr Pierucci. What I tell you is going to make you nervous, but I want you to know that tomorrow morning you will be so glad to see us.'

To be honest, I'm not sure if this is a strand of sadism or a friendly warning. I've never set foot in a prison before. At the entrance, two guards order me to undress. They take everything: my watch, my wedding ring, my shoes. I'm completely naked and so disoriented that I lose my command of English ...

'Turn around, squat and cough,' the guard orders me, in an accent that I can barely understand.

*Cough*, I get that. But *squat*? I don't remember what that means.

'Squat and cough,' repeats the guard angrily. 'Squat and cough!'

Seeing my dumbfounded look, he mimics what I must do. I have to crouch (squat), spread my legs and cough. I comply while the guard stands behind me.

He's checking to see that nothing falls out of my butt.

Squat and cough. Since then, I have retained the expression. I had to comply with this humiliating procedure dozens of times during my incarceration. That evening, as if in a trance, I discovered the US penitentiary system. The guard orders me to put on an orange jumpsuit. I then wait more than two hours, standing, hand-cuffed, with my hands behind my back. The prison has run out of English admission documents. There are some in Spanish or Chinese, but none in English ... Once we get the papers back and fill them out, I'm taken to the cell. In fact, I later find out that they have put me in 'the hole', where the most dangerous prisoners are placed in solitary confinement. It is almost 3 a.m. A guard pushes me inside. I am surrounded by darkness but it's not totally black. No, it's grey. A tiny neon light casts a gloomy glare. The guard closes the door. I then realize that I'm still shackled in the back. For the first time, I start to panic. I can feel the anxiety mounting. They're going to leave me handcuffed all night! Suddenly, I hear a banging sound. A small hatch opens in the door, and the guard shouts at me to step backwards. I obey, walk to him in reverse mode and he removes my handcuffs through the hatch.

Ross and Ron were right. The first night in detention is grue-some. The stench of the cell, its suffocating crampedness. I can't see anything, but I can hear. From all directions I hear violence and blood-curdling screams, as if they are fighting and slaying each other. I haven't eaten or drunk since my arrest. It's impossible to sleep. But this detention is merely an interlude. I spend the night trying to remember the facts surrounding the Tarahan contract ten years ago and revising my schedule. Okay, so I have missed my first morning of appointments in Connecticut. It's not the end of the world. I view the pages of my schedule. I just have to reschedule this meeting to late morning and this one to early afternoon. By playing tight, I should be able to complete my entire programme in twenty-four hours instead of forty-eight.

I will be back in Singapore in three days and back home as planned on Friday. Meaning that, this weekend, I will be able to take my young twins (Raphaella and Gabriella, seven years old) to their friend's birthday party, and the older twins (Pierre and Léa, fifteen years old) to their football match. It seems dumb in retrospect, but this thought reassures and relieves me. I doze off for a few minutes.

# 3

# The first hearing

Who would have thought it? In the early morning, I am indeed glad to see the two FBI investigators again. After a further strip search, I am transferred in shackles to the New Haven Court, a two-hour drive from New York. During the journey I feel like I'm returning to an almost normal life. Ron and Ross bring me coffee and bagels, and they are more than happy to chat. They are both thirty-five years old. Ron has three children. He's a tall, beefy guy, passionate about scuba diving. As for Ross, he's the father of a little girl. They are both very keen to discover France. We are now talking as if we were old acquaintances.

When we first arrive at the court, Ron and Ross park outside and await instructions. We are early and therefore wait a good hour, sitting in the car, until my two guards are told finally that the hearing will not be held in New Haven but in Bridgeport, about thirty minutes away in the opposite direction. Off we go again. Before I return to their custody, Ron parks his car and Ross gives me my phone. It means it's my last chance to make a phone call to a family member if anything goes wrong at the hearing. It's noon here, so midnight in Singapore. I decide to call Tim Curran, the head of Alstom's power division in the United States.

I need to update him about the conversation I had with Keith Carr the day before. I haven't forgotten that the latter is due to arrive in Washington later today. It seems obvious to me that Tim Curran will be monitoring the situation with the general counsel. That's what I'm asking him to do at any rate.

I say goodbye to Ross and Ron, who hand over to a marshal, who locks me in a courtroom cell. The hearing to examine my request for release on bail is about to begin. First, I finally have the right to meet with the legal counsel appointed by Alstom. I enter a small box and meet Liz Latif from Day Pitney. A young woman of about thirty-five to forty years of age.

My first impression of her does not fill me with confidence. She is unfamiliar with Alstom's case and does not seem to be familiar at all with the offence I am charged with, i.e. violation of the Foreign Corrupt Practices Act, the extraterritorial law that gives the Department of Justice the right to incarcerate anyone, regardless of nationality, from the moment they are suspected of having committed the offence of bribing a foreign public official, which may be linked in one way or another to the United States' territory. Liz Latif informs me of some facts.

'It was Alstom's lawyers, Mr Pierucci, who contacted our firm this morning to ask us to assume our defence, as they are not allowed to do it themselves.'

'Why is that? It would have made more sense for them to take ownership of my file?'

'Yes, but there is the risk of a conflict of interest . . .'

'I don't understand; just include me in the deal that Alstom is making with the Department of Justice on the Indonesian case: that seems to me the least they can do. Where is the conflict of interest between Alstom and myself?'

'It's not as straightforward as that, Mr Pierucci; however, you should know that Alstom has agreed to pay for your defence.'

I then attempt to obtain from Liz details of the charges against me. It is not easy to converse in the small box reserved for meetings between detainees and lawyers, as we are separated by a wire grille. She tries to show me some papers by pressing them against the wires. Obviously, I can't read them properly. Moreover, I note that she has not had time to read the indictment.

'But what exactly is the accusation against me? Did you at least get to see the main lines of the file?'

'One case of corruption, and money laundering.'

Money laundering! An offence usually attributed to arms traffickers, or drug dealers! Where the hell did they get that wild accusation? Liz, seeing me enraged, tries to reassure me.

'Either way, they are not going to discuss the merits of the case today. I'm simply going to ask for your release. I shall suggest bail of a hundred thousand dollars, which should be ample to convince the court. You should know that you were indicted by a Grand Jury and that your indictment was kept under seal until your arrest. It is no longer confidential, and the DOJ will certainly be contacting the press today. You should also know that you are not the first Alstom person to be prosecuted. One of your former colleagues, when you were based in the United States, David Rothschild, has already been indicted and cross-examined by investigators. He agreed to plead guilty and negotiated a maximum five-year jail sentence.'

Rothschild pleaded guilty and is facing a sentence of five years, maximum! I turn completely white this time. Suddenly I become brutally aware of the gravity of the accusations and especially of the potentially disastrous consequences both for myself and for my family. But I hardly have time to reflect on these before an usher beckons us. The hearing begins. It is presided over by Judge Garfinkel, who after asking me if I understand English well enough, gives the floor to my defence. It takes less than a minute for Liz Latif to argue for my release on bail of $100,000 plus electronic tagging.

Then the floor goes to the prosecuting attorney representing the US government, Novick, who interrogated me at FBI headquarters. And this time, he is a roaring success.

Novick is firmly opposed to my release. He spews out his arguments with fury. And, unashamedly, asserts the exact opposite of what he had told me in the FBI offices.

'Mr Pierucci is a very high-ranking executive of Alstom. The corruption case in which he is implicated is extremely serious. His company paid bribes to an Indonesian government official to obtain favours. We have built up a solid prosecution case. As well as a great deal of documentation, we have evidence of his involvement in a conspiracy to violate the US Anti-Corruption law of the Foreign Corrupt Practices Act.'

So this is the price I have to pay for refusing his deal in the first interview. Then Novick starts telling the court that I am a flight risk.

'Frédéric Pierucci has no ties to the United States. When he worked here, he was issued a green card [a permanent residence permit]. Yet he suspiciously returned it to the authorities in 2012. We interviewed the employee to whom he had entrusted this document. The latter told us that he found Frédéric Pierucci's behaviour rather strange.'

I must be hallucinating. In 2012, during one of my many trips to the United States, I simply decided to return the green card, as I no longer required it. A few weeks earlier, I had been transferred to Singapore. How can this be deemed suspicious? However, Novick was not finished yet.

'If you release him, he'll most certainly flee. And as you know very well, Your Honour, France does not extradite its citizens. Furthermore, he was charged and an arrest warrant was issued, but he still did not surrender to the authorities.'

I find this attorney's line of argument quite staggering. How could I have surrendered myself to the authorities when I had no idea that an arrest warrant had been issued against me, since the DOJ had kept this information sealed until today, considering me a flight risk? If I had known, I would probably have sought legal advice before travelling to the United States on business. It's quite simply absurd. Nevertheless, Judge Garfinkel seems to be troubled and I hear her say: 'I must say that the government has presented a very strong case. The defence needs to constitute a stronger probation case if I am to release her client. Ms Latif, I would like to give you some time to prepare a new motion. When do you think you will be ready?'

'Late afternoon, Your Honour.'

'That's not possible, I unfortunately have to leave in an hour for a medical appointment. I suggest we meet again in two days.'

The hearing is adjourned. The judge then turns towards me and says: 'How are you pleading, Mr Pierucci? Guilty or not guilty?'

'Not guilty.'

I was only asked one question. And I was only allowed to utter one or two words. I just have time to figure out that I'm going to stay another forty-eight hours in jail before returning, still hand-cuffed behind my back, to a court cell where they give me two minutes with my lawyer. I urge her to inform Keith Carr of the truly alarming twist that my case is taking.

Two hours later, the guards pull me out of my cell and chain me up like a beast.

Yes, I have become a beast. There is no other way to portray me: my wrists and ankles are shackled by handcuffs, my upper body is restrained by a heavy chain, the handcuffs and chain attached to a huge padlock that rests on my stomach. The only time I ever saw human beings shackled in this way was on television, during reports on prisoners in Guantanamo Bay. As I can no longer walk normally with these chains locking my ankles, the guards force me to jump with my feet together to reach an armoured vehicle waiting for us in the courthouse basement. The black vehicle, with its armoured windows covered with thick wire mesh, resembles a special forces intervention truck.

There are two other detainees sitting next to me. An Asian and a big black guy. I try to start a conversation: 'Do you know where we are going?' I cannot understand a single word they say. They are using jail vernacular, a kind of backward slang with coded expressions.

Beaten by exhaustion, I don't press them further. I haven't slept for almost two days now. I am literally numb, stunned by the sequence of events. In this armoured truck, this rolling-stock jail, trussed up like wild game, the fatigue knocks me out and I fall asleep. Five hours later, I wake up at Donald W. Wyatt Detention Facility, in Rhode Island.

# 4

## Wyatt

How can I describe the Wyatt Detention Facility? Seen from afar, or from overhead, this prison resembles an ordinary five-storey administrative building, no different from the surrounding buildings. But as you approach, you discover that it is a genuine blockhouse, a concrete sarcophagus, with tiny slits in the facade in place of windows, each measuring six inches wide and thirty inches high. Loopholes that send shivers down your spine, and make you wonder how daylight can pass through. You feel that once you're in there, anything could happen. Wyatt is cut off from the rest of the world. Surrounded by a double fence, a barbed-wire field, and surveillance cameras every ten yards. All vehicles entering it are armoured. Wyatt is no ordinary prison. It's a maximum-security detention facility.

Americans use a scale of one to four to rate the level of security of their jails. Level 1 establishments, known as 'camps', are usually reserved for white-collar criminals convicted of financial crimes. These camps are equipped with gyms and often tennis courts, few guards and minimum surveillance measures. Security 2 centres are for short sentences and non-violent prisoners. Then there are the so-called mid-level detention centres, classified as level 3, and finally maximum-security establishments.

Wyatt belongs to the latter category. This prison is home to the most dangerous criminals in Connecticut, Massachusetts, Rhode Island, Maine and Vermont. They are detained there awaiting trial. Wyatt is therefore not part of the Bureau of Prisons, which groups together the federal penitentiaries where prisoners who have already been tried are incarcerated. Wyatt is managed by a private company under the authority of the Bureau of Prisons. The detention facility houses an average of six hundred inmates, who, as is customary practice in the United States, are dispersed into *pods* ('quarters') according to various criteria such as gang membership, age, dangerousness, ethnic origin, etc. In 2013, according to the annual report of the Wyatt administration, this figure included 39 per cent Hispanics, 36 per cent African Americans and 25 per cent Caucasian Whites. The same report also points out that, in 2013, several cases of sexual abuse among detainees were reported but not addressed. Also, during this same period, two detainees were murdered in such horrific circumstances that the families of the victims decided to file a suit.

So it is in this ultra-secure facility that the Department of Justice has decided to place me. Yet I am neither a re-offender nor a dangerous inmate. Such a choice defies any custodial logic, but nobody deigns to provide me with the slightest explanation.

On 15 April 2013, when our convoy passes through the gates, we are blocked by a first security airlock, before a grid rises and we proceed to a second airlock. Here, they pull me out of the van along with the other two passengers whose lingua franca I still can't decipher. Having to jump, as our ankles are shackled, we pass through three armoured doors in succession to finally reach the R&D (Receive and Discharge) room, the building that manages the in-and-out movement of prisoners.

This room features a counter, behind which sits the steward who is responsible for receiving arrivals, a security screening gate to detect metals, similar to those seen at airports, two booths for body searches, and a special chair used to restrain the most violent prisoners. The guards remove our handcuffs. And once again we have to strip naked. This is my fourth body search since my arrest and as I haven't washed since I left Singapore – two days ago – I must reek. But strangely enough, I don't care. It has taken only forty-eight hours for me to start losing my most elementary bearings. It's all a haze. I float, as if I were moving into another dimension . . .

I barely react when the prison staff hand us our kit. At Wyatt, new arrivals wear a khaki uniform as in all American federal penitentiaries, except when you are in 'the hole', where the colour is orange. We are also allowed four boxers, four pairs of socks, four T-shirts, two pairs of trousers, a pair of sneakers, and a pair of flip flops. Apart from the shoes, everything is old and shabby from having been worn before. The guards also hand me a badge featuring my mugshot, which they took in front of a grid indicating my height, just like in the film *The Usual Suspects*. It bears the number 21613.

Next, we must complete the admission questionnaire, which includes a list of contacts to be mentioned with their phone numbers. I suddenly realize that I don't know any of the phone numbers of my nearest and dearest by heart, not even Clara's in Singapore, which has just changed. I have no means of contacting my lawyer either. I start to panic. Liz Latif hadn't thought to leave me her contact details. The only American 'official' I can call, because he had the foresight to leave me his business card, is Seth Blum, the investigator who received me at FBI headquarters. I must reach him, at all costs, to let him know where I am.

'No way,' mutters the guard, a Hispanic with an emaciated face. I insist. I try to explain the situation to my jailer, which annoys him even more. He locks me in a cell with the two other passengers from the van, then returns at the end of an hour. Only God knows why, but he changed his mind. I can make a call, but only one.

I pray that Seth will pick up, which he does. My luck ends there. He's on the train taking him from New York to Washington, and before he has time to give me Liz's number, the line goes dead. I just had time to explain my problem to him. Naturally, I ask the guard to call back.

'This is not a hotel, asshole! You were told one call, not two! Get the hell outta here!'

I try explaining, I almost beg . . . Nothing works.

'One phone call! And if you carry on acting smart, I'll put you in the hole!' screams the guard. But this vicious-looking screw is not letting me argue, so I have to accept it.

Before leaving the 'intake room' to join the pod allocated to him, each inmate is assigned a toothbrush and a small tube of toothpaste, soap, a small bottle of shampoo, two towels, a two-inch-thick plastic mattress, a pair of sheets and a brown blanket. I'm in D pod, one of the most dilapidated in the prison. In Wyatt, the pods are organized around a common room, surrounded by cells. D pod has about twenty cells. Each cell can house four inmates. For the moment, I share cell number 19 with my two travelling companions. We'd better get along because the prison's internal regulations stipulate that during the first seventy-two hours of detention, we are not allowed to leave our cell, except to go to breakfast, lunch and dinner at 7:50 a.m., 12:20 p.m. and 5:20 p.m. respectively. Apart from walking through the common room, which doubles up as a refectory, all three of us have to remain locked up for almost twenty-two hours a day in a space of thirteen square yards during this first stage of the incarceration process.

The cell is equipped with a small iron table, a sink, a toilet, two stools fixed to the floor and two bunk beds. The cells were designed to accommodate two inmates, but due to overcrowding, they now host four. There are no partitions separating the toilets. The only way to get some privacy to relieve yourself is to wait for the guards to activate the automated cell door opening at mealtimes. This enables your cell mates to wait outside in the hallway for a few minutes and give you a break.

The Asian guy settles on the bunk above mine, and the tall black guy slumps down opposite me. Fortunately, my cell mates are pretty good company. They grasped that I can't understand anything they are saying, so they talk to me more slowly, paying attention to their words. We pass the time by sharing our respective stories. Cho has an odd tale to tell. He is a Vietnamese political refugee who, after a hellish time in transit camps in Malaysia, managed to emigrate to San Francisco in 1991. With his meagre savings, he opened a first restaurant, then a second, and ended up making a fortune in the catering industry.

'I managed to save two million dollars,' he tells me. 'And then I screwed up. I was a casino freak. I lost everything, so to bail myself out I started making fake credit cards.'

Cho was arrested on one occasion and given a two-year custodial sentence, which he served in California. Once released, he relapsed, and gambled away the astronomical sum of $12 million! He was arrested again for committing large-scale fraud, and now faces a ten-year sentence.

Mason, on the other hand, has a more 'classic' background. He grew up in the black district of Hartford, the capital of Connecticut. Unknown father, drug-addict mother. Mason was only fourteen years old when he joined a gang and started trafficking cocaine out of Texas. He spent six years behind bars, and upon release, he became a member of the '666', a Muslim sect reserved for Blacks and openly racist towards Whites, which boasts of enforcing its own law even within prisons. He was subsequently sentenced to another eight years' imprisonment. However, between two incarcerations, he succeeded in producing four children from four different women in two years. He proudly explains that these are four 'very fine' ladies.

'One is even a prison guard! The second one works in a museum security department; the third is a waitress at McDonald's; and the fourth is a stripper in a Hartford club.' He then hilariously adds, 'And you know what? None of them have tried to claim alimony from me.'

On that first day in Wyatt, my cell mates also familiarized me with prison codes. As I leant over the sink to brush my teeth and spit into the sink, Mason started yelling and insulting me.

'You cannot spit. You cannot do that. You gotta do that in the john. You cannot spit where we all wash!'

It soon became apparent that the prisoners are very strict on hygiene issues.

'It's the same when you piss. You have to sit down and piss like a girl,' says Mason. 'You got it? You mustn't spill it everywhere. You must never piss standing up. And if you want to pass wind, it is the same. You do it on the john and flush it so it sucks up the smell, you understand?'

Message received loud and clear. Indeed, all these rules make sense. I learn them in stages. My inmates know from experience that if one of us gets ill in the cell, the risk of infection is very high. Medical aid at Wyatt is almost non-existent. This is something I discover very soon for myself.

Just before I flew to New York, during my last tennis match, the first one for a long time, I suffered a serious rupture of the external and internal ligaments in my right ankle. I got on the plane barely able to walk (you can imagine how it felt when I had to jump with my ankles shackled). Upon my arrival in Wyatt, despite my repeated requests, I received no medical care, apart from an aspirin.

Even though Cho and Mason are naturally quite sociable, these first few hours of detention seem endless. No music, no TV, no notebook, no pen, no book. The only document I was able to keep was a summary of the indictment that Liz handed me in court. Reading it, I revert to the early 2000s, when the goddamn Indonesian contract was negotiated; the reason I am locked up in this pen today.

# 5

# Recollections

Ironically, at that time in my life, I was thinking of leaving the company. I was thirty-one years old, and after four years in Beijing (1995–99) as Sales Director China for the Power division, I was keen to make a career change. Indeed, since I joined Alstom, I had made significant career strides, but as a graduate of an average engineering school (ENSMA in Poitiers), I was afraid of hitting a glass ceiling fast. I knew that in order to progress in a multinational I would have to obtain further qualifications, so decided to leave and do an MBA at INSEAD, one of the world's leading graduate business schools, which offered me a place.

In 1999 Clara and I discussed this at length. After agreeing to put her career plans on hold to follow me to Beijing, then giving birth to our first set of twins, Pierre and Léa in January 1998, and completing her PhD in neurobiology, she was keen to return to work. Hence our desire to settle in France.

Today, with hindsight, I bitterly regret that I did not stick to this choice. Who knows what the future would have held for us, or even if we would have been happier, but what I do know is that I would never have ended up here, in penal servitude.

Alstom, at that time, knew how to hold on to me. They saw me as young rising talent. After China, I was offered a senior role in the United States as Vice President Global Sales and Marketing of the Boiler division. To persuade me once and for all, my managers offered me time off (every other Friday plus several weeks a year) to attend the MBA courses at Columbia University in New York, one of the most prestigious American universities, a member of the famous 'Ivy League', and they agreed to fully fund my tuition for an amount of $100,000. Who could refuse such an offer?

In September 1999, I left for Windsor, Connecticut, where Clara and the children joined me two months later. Soon after my arrival, however, my mission proved to be much more challenging than expected.

At the beginning of the year 2000, Alstom was financially distressed. It was on the verge of bankruptcy. One year prior to this, management had made an alliance with ABB, a Swiss–Swedish competitor, but very soon this industrial partnership was to prove catastrophic. Whereas Alstom thought it had made the deal of the century by seizing control of ABB's gas turbine technology, sold and distributed all over the world, it had just signed the worst deal in its history. These gas turbines were poorly designed and suffered from numerous technical faults. As a result, Alstom had to pay over €2 billion in compensation to its clients, which in turn increased its debt by an unprecedented 2,000 per cent. It posted a record deficit of €5.3 billion, which gave lenders a fright.

At this point, the board of directors decided to dismiss Pierre Bilger and hand over the reins of the group to Patrick Kron in an attempt to redress the situation. This choice was well received in-house, for Kron belongs to the 'elite', or rather the elite of the elite. He is an 'X-Mine' (alumnus of the elite French university for top scientists), selected from the list that each year assembles the top twenty students from the Polytechnics and the École des Mines. They represent a small aristocracy, or even a republican oligarchy, which for two centuries has been at the helm of France's biggest companies and economy. After having spent the first part of his career at Péchiney, he became a director of Alstom in 2001, then Chief Executive Officer in January 2003, and ultimately Chairman and Chief Executive Officer. In the months following his ascension to power, he set about rescuing the company. To avoid bankruptcy proceedings, he even pleaded in person before the Paris Commercial Court, as well as in Brussels before the European Commission, and convinced the state to bail out Alstom, in exchange for a 'realignment' of the company's activities and a major overhaul of its management. Over two hundred senior managers were given their marching orders. Throughout this salvage operation, Patrick Kron had the support of Nicolas Sarkozy. The future President of the Republic, at the time Minister of the Economy, knew how attached the French were to their large enterprises and did not want to appear as having stood by idly watching the demise of one of our rare multinationals. He succeeded in obtaining a partial renationalization of the company, with the French state acquiring just over 20 per cent of the capital. With the help of Patrick Kron, Nicolas Sarkozy became Alstom's saviour.

At the time, I was light years away from the battles being fought at Alstom's headquarters or within the government.

On arrival in the United States, I found myself in a real 'vipers' nest'. The Windsor, Connecticut unit where I was based was run entirely by ABB, with whom we had merged at the end of 1999. And to cap it all, I found myself having to deal directly with a senior manager, who disliked me.

In fact, a year earlier in 1998, while we were still rivals (he was working for ABB and I represented Alstom), we both fought hard to win the contract for the largest power plant in China at the time.

This contract was a priority target for boiler manufacturers around the world. The final decision was between ABB and Alstom, and Alstom walked away with the prize. This explains the wrath of my new Windsor colleague, who had missed his chance to take over global management of the boiler division. This position was eventually assigned to another former ABB executive.

Our relationship did not improve either when headquarters asked us to provide them with a complete list (and copies of contracts) of all the 'consultants' worldwide with whom we had contracts.

Before France's ratification of the OECD Anti-Bribery Convention in September 2000, the use of intermediaries, 'consultants', to obtain international contracts was a perfectly tolerated practice. While the use of bribes on French territory was strictly prohibited, it was permitted abroad. Therefore, each year, the heads of French corporations would go to the Finance Ministry in Bercy to compile a list of their 'exceptional expenses'; in short, the bribes paid to intermediaries or consultants, to win international contracts. These amounts were duly identified and then deducted from corporation taxes. It was the state's very pragmatic way of legalizing the illegal.

However, after September 2000, this changed. France, like other countries before it, was committed to combating international corruption. As a result, Alstom's management asked for a comprehensive insight into the undertakings made by ABB with all its intermediaries in order to comply with the new French legislation. My boss assigned this delicate task to me. I easily obtained the names and contracts of the consultants employed by the boiler business units from Alstom. However, it was another story for those from the former ABB (including the American unit in Windsor). Although we had merged, these units were reluctant to cooperate and disclose their networks of intermediaries. Furthermore, in each country, ABB companies behaved like local barons and did not want to bow to centralized control. Nevertheless, I managed to establish a first inventory. I ended up with a pile of contracts on my desk, each one drafted with different terms and often containing truly staggering clauses. Some consultants had succeeded in negotiating agreements with no expiry date, sometimes even with monthly payments, i.e. a licence to long-term corruption.

In the meantime, Alstom's management started introducing new processes to show its commitment to the tightening up of its compliance processes (compliance with norms and laws and business ethics). Henceforth, a very specific procedure for the approval of intermediaries was implemented. As many as thirteen signatures would be required before a consultant could be hired. At the end of this process, a 'project sheet' must be established for each contract, which must clearly indicate the amount of the commission and payment terms (time frame and payment schedule). This sheet must be approved and signed at the highest level by three people:

1. The Senior Vice-President of the division preparing the offer for the project.
2. The Senior Vice-President in charge of Alstom's International Network.
3. The Regional Senior Vice-President of the International Network region where the project is based.

Finally, all projects worth more than $50 million – i.e. almost all bids for the boiler business – become subject to validation by the 'risk committee', which reports directly to the CEO and includes the Chief Financial Officer.

In addition, the group set up a specific company within its structure, 'Alstom Prom', based in Switzerland, in charge of drafting, negotiating and signing virtually all the contracts concluded with consultants. This company was at that time headed by Alstom's compliance officer, the same person responsible for enforcing the law and ethics within the company.

But don't be fooled, this amounted to window dressing. The processes implemented since 2002–03 were purely cosmetic. There was never any question of launching a clean-up operation within the group. The only way to curb corruption would be to cease the use of consultants once and for all. That is not what was decided. In fact, it was quite the reverse. Under the guise of 'project sheets' and more rigorous 'processes', the use of consultants was largely continued under Patrick Kron's leadership. The only difference being that, henceforth, the corruption became more covert.

On the surface, Alstom strictly complied with all the rules. For instance, all contracts featured two clauses: one detailing the anti-corruption law in force, and the other reminding the consultants of their obligation not to pay bribes. Such clauses were deemed by law practitioners as security in the event of prosecution. But behind this facade of good repute, Alstom continued to remunerate consultants to exert influence on ministries, political parties, consulting engineers, experts and evaluation committees in numerous countries. If the risk appeared too great, the group, rather than using consultants directly, would use the services of local subcontractors (civil engineering companies, erection companies, etc.), who were less hampered by the anti-corruption provisions in place. Alstom was by no means an exception. Many multinationals, all of them advised by the same major international consultancy firms, adopted this type of procedure.

The compliance department responsible for ensuring compliance with a code of professional conduct of course knew about these practices, as did the group's senior management. And for good reason: they had set them up!

# 6

## The phone call

First night at Wyatt. Very agitated. I barely closed my eyes. Mason snores like a locomotive. The call to breakfast is almost liberating. I can finally exit the cell and take a shower. As the first to arrive in the communal shower room, I undress and start to wash, soon joined by another inmate who immediately hurls abuse at me.

'We do not shower naked here! You keep your boxers and flip flops on. Not to contaminate the place.'

I have a lot to learn. And fast. At Wyatt, almost all inmates are repeat offenders. Everyone knows the rules of prison etiquette. I am the only 'rookie'. And I risk having my head served up if I don't take a crash course ASAP.

I also desperately need to find a way to contact my elusive lawyer, Liz. Once more I beg the Correction Officer who is monitoring the pod for permission to phone Seth Blum. 'Take it up with the social worker,' he replies. 'She'll be here after lunch.'

I'm having to learn patience, a cardinal virtue behind bars. The social worker shows up right after lunch, as promised. But it's a mad rush. They all scramble to talk to her. Wait and wait ... Prison is first and foremost about waiting. Finally, it's my turn. The social worker receives me and I discover in disbelief that, notwithstanding the fuzzy communication during my desperate call to Seth Blum, the FBI inspector figured out what I wanted and, even more extraordinarily, he called Wyatt back and insisted that the supervisors give me Liz Latif's number via the social worker. It makes me feel that good old Seth cares as much about my defence as does my own counsel.

But before I can reach her, I have a new hurdle to cross. Accessing one of the four wall phones installed in the common room, after observing a very exacting procedure. As a newcomer, I am required to provide Wyatt's management, via the social worker, with a list of the people I will be contacting during my 'stay'. Such a list has to be duly approved and registered by the prison authorities.

Problem number 1: apart from Liz's number, which I was just given, I don't know any other number by heart. Problem number 2: you have to pay before you can make a call. Each prisoner has to have a credited canteen account from which the exorbitant cost of telephone communications is debited. But my wallet and credit cards were confiscated, like all my other belongings, and entrusted to Liz! This is akin to Kafka in Connecticut! The social worker agrees. And she 'exceptionally' authorizes me to call my lawyer on her office line.

So, I can finally talk to Ms Latif again, who discovers in what prison I am housed.

'But where are you?' she naively enquires.

I had imagined her frantically trying to work out where I'd been imprisoned, but apparently not. The remainder of our conversation is no more reassuring either.

'Well, Mr Pierucci, things are not looking very good . . . The hundred thousand dollars' bail, which I envisaged offering the court in exchange for your release, is deemed very inadequate. Clearly, the Department of Justice wants you to remain in detention. The US attorney is going to raise the stakes. How much do you have in your bank account?'

I quickly do the maths in my head.

'Putting everything together, maybe around four hundred thousand dollars.'

'Hmm. That might still be a little tight. You cannot do more?'

'No, I may be a senior executive, but I'm not that wealthy. I own a house in the Paris suburbs that was purchased on 100 per cent credit, and that's all. But I'm not on my own in this, am I? What about Alstom? I'm their problem too. I guess they're going to step in?'

'No doubt. Well, look, I finally got a new bail hearing for tomorrow morning. So we'll know soon enough. Don't worry too much. We'll find a solution.'

'I hope so. And please also pass the message on to Keith Carr, Alstom's general counsel, that I want him to come and see me here in Wyatt as soon as he has finished his meeting with the DOJ in Washington.'

Our conversation is now over. She had time to look in my BlackBerry and give me the phone numbers of my wife Clara, my sister and my parents. I gave her my credit card code and asked her to deposit $50 to my canteen account at Wyatt ASAP in case I need it . . . As usual, especially in the most challenging situations, I try to reason as coldly as possible. Is this due to my engineering background or my liking for mathematics? I tackle complex situations as operations.

The good news is, I only have a few more hours to survive in this hellhole of a prison. Tomorrow morning, the judge should order my release, even if I have to put my house up for bail. Any other decision seems unimaginable to me in a country that releases murder suspects onto the streets. The worst news is that the judge responsible for deciding my fate could, in exchange for my release, prohibit me from leaving US territory while awaiting trial. In terms of family and work, this situation is of course far from ideal. But it is not catastrophic either. I already worked in Connecticut for seven years, before leaving for France in 2006. I know the American subsidiary perfectly well, so I should be able to continue to run the boiler business from the United States instead of Singapore without any difficulty, at least for a few months. On the understanding, of course, that my company grants me customized status. Given all the trouble the latter has caused me, it should be accommodating, at the very least.

In terms of family, however, this could become a nightmare. With Clara and our four children, we only recently moved to Singapore. We arrived in August 2012. This move to Asia was a godsend for the whole family. After overcoming some difficulties in our relationship, the move to Singapore heralded a new start. And the wager was won. The children were happy there. Our two sets of twins settled in perfectly. They liked the new international schools they attended and quickly made many friends. I can still remember Gabriella, the first week after our arrival, walking through the rooms of our large house with her iPad, giving her grandfather a guided tour. She was so proud and happy, as were her sisters and brother. And Clara and I gradually reconnected.

It has now been almost seventy-two hours since I gave her any news. Although I rarely call when travelling, she will need to be informed. I'll tell her tomorrow after the hearing. That is what was arranged with my lawyer. By then I will be free again and the impact won't be as harsh. So how am I going to explain to her what's happened to me? If I have to remain in the United States for several months, while awaiting trial, then how exactly are we going to organize things? Will the whole family have to move again? Thanks to her PhD in neurobiology and professional experience, Clara has just been recruited by a major French company based in Singapore. She loves her new job. Maybe it would be better if I moved to Boston by myself for a while? But could Clara handle such a separation? And the children?

I lie on my bunk in the cell, mulling over my thoughts. The questions go round in my head. I reread the summary of the indictment and continue to fill in the timeline of events surrounding the Tarahan case, but it was all so long ago . . . The hearing is scheduled for tomorrow morning at 11 a.m. Therefore, calculating the time it takes to get me out of the cell and ready, the guards will come to wake me before daybreak, i.e. 4 a.m. I should really try sleeping, but the bunks are so narrow, only twenty inches wide, and the plastic mattress is so thin that I worry about falling out of bed in my sleep. My fellow inmates did show me a technique to avoid falls: you have to package the mattress, blanket and sheet using large knots. It works, but you feel so tightly squeezed, like a perfectly tied-up roast. I don't get a wink of sleep. Motionless and silent, I wait.

# 7

## They forgot me!

As soon as I arrived in Wyatt, the supervisors confiscated everything from me. They took my watch, my wedding ring, and I have lost all notion of time. It is daybreak, a ray of light is slowly breaking through the slit that serves as a window in the cell. I'm still waiting. I listen out for the slightest noise with the hope, each time, that the guards will come and get me to escort me to the courtroom. No one comes. It must be at least 6 a.m. What if they have forgotten me? I tap on the door. Nothing. I tap again and again. Louder and louder. Finally, a warden agrees to talk to me. This time it's not unwillingness that I detect in his face, but sheer surprise. He swears to me that neither he nor any of his colleagues received any instructions to arrange for my transportation to the New Haven courthouse. But he checks it out.

When he returns, he confirms that I definitely am not on the agenda for extraction. I'm devastated. I feel like I'm going crazy, trying not to let paranoia get the better of me. Suppose my lawyer lied to me? What if she's in collusion with the prosecuting attorney? After all, I don't know anything about her, it was Alstom who chose her. How can I trust her? I have never felt so helpless. I punch the door again with my fists. The screw puts his head in the cell, but has run out of patience.

He gets angry as I continue to argue with such frenzy. I explain to him that it is vital that I call my lawyer, that all this is a huge misunderstanding, that I must get out, that I am being summoned by a judge who will release me, that all this is absurd and that he must help me. The guard leaves and reappears one minute later with a book to help cure my stress.

Wyatt Prison's internal regulations. This fifty-page book stipulates, inter alia, the conditions under which a prisoner may file complaints to the prison administration. I want to scream. What do they want from me? To drive me crazy? Put me in a straitjacket? Then little by little I calm down. Besides, I have no choice but to shut up and wait a very long time. It was only in the middle of the afternoon that I managed to call Liz Latif.

'The prison officers at Wyatt screwed up big time,' she said. 'They simply forgot to come and get you from your cell. The hearing that was supposed to decide on your bail application began as scheduled; however, the judge, after noticing your absence, decided to defer it again for another two days!'

Do not crack. Take a deep breath and remain focused.

'Under these circumstances Liz, it is imperative that you inform my wife. She must be worried sick.'

'I'll do it right away. Don't worry, Mr Pierucci. I'm coming back tomorrow with my boss. I shall also bring the key exhibits from the prosecution file with me. We will have time to examine them.'

Finally, I shall find out what exactly I am accused of, because reading the summary of the indictment that she sent me forty-eight hours ago did not provide me with all the necessary information and prompted more questions than answers! Our conversation ends. I am escorted back to my cell with the grim prospect of having to remain locked up for hours on end. For someone hyperactive like me, this is difficult to bear. I have absolutely nothing to do. So, to kill time, I read over and over again the *Prison Inmate User Manual*. It provides an extremely detailed description of prison life. The chapter entitled 'Contacts with the outside world' covers several pages. I now understand more clearly why the social worker ended up letting me use her work phone on an exceptional basis, as if this were a huge privilege. The procedure governing external communications seems to have been drafted by the CIA. Not only must a list of telephone numbers be sent to the prison administration, which decides whether or not to validate it, but the people who have been authorized to contact an inmate must register on an internet platform, from a bank account opened in the United States. Well, this is a real headache for a foreign national. In short, this procedure takes at least a fortnight . . . and is costly.

Like everything else! Because nothing is free inside this prison. Even an inmate's most basic and essential daily necessities such as soap, toothpaste, toothbrush, shower sandals, including his plastic glass, the most prized object in Wyatt because, for a reason that remains unfathomable to me to this day, drinking water is only available in the form of ice cubes. So when you want to drink, you start by leaving your cell (if permitted), which overlooks the common room, to get your (free) ice cubes, arranged in a single tray. Then you must place them in your (payable) plastic beaker that you previously ordered from the prison canteen. No other containers are tolerated or available. Once your plastic beaker is filled with ice, you must go to the only boiling water dispenser in the room to fill your glass. When there is no more ice, which happens all the time, you must wait for delivery of the new tray, once a day. You're lucky if you succeed in bringing your glass of drinking water back to your cell when you need to quench your thirst. Like all things in jail, this ice-melting process is only allowed at certain times of the day.

The common room, which I soon discover also serves as a refectory, is the hub and only living space of D pod. Meals are served in brown plastic trays divided into four compartments. The first one contains two slices of bread, the second green vegetables – though it is often empty – the third is for the main course: a kind of mush that varies in colour depending on the day, and whose content is indefinable. No flavour and, weirder still, no smell. It's impossible to tell what we're eating. And the last compartment is meant for desserts, but the only one ever served is apple puree. Wyatt is a privately run detention facility.

The cost of meals is calculated to the nearest cent. Each meal must not exceed $1 (€0.80). A private institution implies an enterprise, which means profits. Not only should prisoners not cost the local authorities a single dime, but they should also generate revenue for the companies that run the detention centres. Nothing is random. For example, while you can watch TV for free, you must pay to listen to it! For this service, you need to buy a radio and headphones from the prison commissary. Perpetually making payments, this is prison life in the United States.

There are three TV sets fixed to the walls of the common room. One in each corner. One is reserved for the Blacks, showing a unique selection of appalling reality shows such as *Love & Hip Hop: Miami* that feature silicone pin-ups all day long. If you are Hispanic, you will gather around another screen that exclusively spills out Mexican soap operas from the Telemundo channel and sometimes soccer. The last TV set, that of the Whites, continuously diffuses basketball, American football, or 'no limit' martial arts combats, except in the early morning when CNN news is screened for one hour. In principle, everyone is free to sit in front of the screen of their choice, but the 'best seats', i.e. those just in front of the screen, are reserved and automatically occupied by the members of the ethnic group to whom the set 'belongs'. And you must not ask to change channels if you are watching images outside of your ethnic community. It is the guards who supervise the use of the remote controls, as disputes are frequent and sometimes violent.

For the same reasons, the common room is likewise permanently monitored by three security cameras. In front of the four touch-tone telephone sets in the refectory, there are permanent and endless queues. There is no privacy. Everyone hears each other's conversations (restricted to twenty minutes maximum). Not to mention that each call is listened to and recorded by Wyatt's authorities, and then forwarded to prosecuting attorneys and FBI investigators. Finally, adjoining the same common room are the communal showers, two of which are out of order. As I'd learned previously, the prisoners shower dressed in their flip flops and boxers, as much for the sake of hygiene as to avoid sexual assault. Welcome to Wyatt!

# 8

## Stan

'Hello, I am Stan Twardy, former Attorney General of Connecticut.'

This is how my new lawyer introduces himself. Tall, grey hair, sixty-two years old, a Hollywood smile and endless honorary titles. This time they got it right. Alstom has finally sent me someone tough, serious and confident. A perfect match for the challenges at stake. Stan Twardy defends several companies belonging to the club of the 500 largest American firms, and he is the author of half a dozen legal works that have earned him the honour of being selected as one of the 'Best Lawyers' in the United States.

'I shall explain,' he told me in his opening remarks. 'My firm Day Pitney was mandated by Patton Boggs' lawyers, who are defending Alstom in the corruption proceedings. Your firm is paying all of our firm's fees.'

I rejoice at his words. Such a difference compared to Liz, who stands by his side, speechless and seemingly in awe. Stan has perfect self-control. He has a confident tone, his vocabulary is very precise. He gets straight to the point.

'Your firm undertakes to pay for your defence, but if you are convicted, then Alstom may ask you to reimburse these costs.'

Did I hear that right? Am I dreaming?

Stan continues, unabashed; Liz remains silent at his side.

'The reality is, you may well have to pay it back. Whether you choose to go to trial and lose, or if you decide to quit before and accept to plead guilty.'

Yes, I did hear right. It only took me a few seconds to realize that I was indeed awake. I don't need to pinch myself any more. And I explode: 'All this is simply disgraceful and unacceptable! Whatever I did, was done in the name of my company, in strict compliance with all internal rules and procedures.'

'Acceptable or not, these are the conditions that Alstom has asked us to submit to you in order to defend you.'

I can hardly believe such impudence. And once again, I hope that I have misunderstood.

'Do you realize what you're telling me? Alstom is currently in negotiations with the American authorities. Alstom will likely have to admit its guilt and negotiate a monetary penalty. And if I go down the same path, you're telling me that I'll be on my own, that Alstom will abandon me! It doesn't make any sense . . .'

'What makes sense, Mr Pierucci, is that if you belonged to an American group, they would have already fired you!'

He takes the moral high ground, treating me like a criminal. I bear in mind that the only person who can get me out of Wyatt is Stan, so I swallow my pride and lower my tone. He takes out a paper prepared by Alstom, hands me a pen and urges me to sign it. I refuse outright.

'I first want to discuss it with Keith Carr; he must still be in the United States, so he should come and visit me in Wyatt.'

Stan agrees to pass on the message, and despite my refusal to sign his document, continues our conversation. He is now getting to the most important subject: my release.

'Just to be clear, Mr Pierucci. Since your arrest, the stakes have been steadily rising. Liz Latif and I are now of the view that you will have to put a very substantial amount of money on the table if you want to get out.'

I swallow and ask the fateful question, which I discover, the longer I'm here, is the only question worth asking in the United States.

'How much?'

'Alstom agrees to provide one and a half million dollars, and we estimate that if you could post bail of four hundred thousand dollars, that should be sufficient. In addition, your company agrees to rent an apartment and pay two guards, who will monitor you so that you do not abscond and flee to France.'

'Did you say guards? Twenty-four/seven to watch over me and my family?'

'That's it. These were the conditions imposed on Dominique Strauss-Kahn during his criminal proceedings in New York. That said, don't be under any illusions; even if a judge accepts our offer, it will take us a while to get the bail together, to rent you an apartment, and to hire guards, so you won't be able to leave for two or three weeks, even if it all goes smoothly.'

If it all goes smoothly! Does this man have any idea what I am going through? I'm in jail, for God's sake, one of the most dangerous in the United States. Two or three weeks. He is talking to me as if it were a matter of solving a simple administrative problem, a common inconvenience, a small incident along the way. What about Alstom's management? Does he believe that Alstom will allow one of its senior executives to languish in jail without intervening? They are not paying Stan for nothing. They are bound to be tracking his every move. They're exerting pressure on him. The lawyer interrupts my thoughts.

'Please be aware, Mr Pierucci, that we have no direct contact with Alstom. We are not permitted to do so. We cannot talk to your managers. The Department of Justice fears that your company may put pressure on you. We only have the right to interact with our colleagues at Patton Boggs, who are representing your firm's interests and who have mandated us to defend you.'

The image that comes to mind is that of a never-ending slide. Every time Stan opens his mouth, the ground slips further away under my feet. If he has no contact with Alstom, how am I supposed to defend myself? How will I be able to obtain the company's internal documents and evidence proving my good faith? What judge would be able to examine the role I had or did not have in the corruption case leading to the charges against Alstom? The only way I can make sense of this is to conclude that my lawyer has probably not adequately evaluated the complexity of the situation. He may well hold numerous diplomas, but he could sure do with a refresher course.

'Stan,' I say meekly, 'I am accused of having knowledge that Alstom used consultants to obtain contracts. However, I did not make the decision to recruit these external consultants. Alstom has its own highly codified internal processes. Instructions came from the top.'

'Mr Pierucci.' He quickly interrupts me. 'It is too early to go into all this detail. The most important thing for now is how best to prepare your application for release on bail.'

'But how can you plead my case if you do not explain to the judge that I was far down the chain of command, that I was not initiating or deciding the use of consultants, nor found the consultants, nor even finally approved of their selection. Thirteen signatures were required to hire any consultants, and two of the three ultimate signatories both reported directly to Alstom's CEO, Patrick Kron. It is imperative that my company provide you with these documents, especially for the consultants appointed for the Tarahan case. You must demand these documents today.'

While I continue to explain in detail to Stan Twardy the predominant role of several group leaders who were under the direct authority of the CEO Patrick Kron, I notice that he is not taking any notes. He just looks at me, with a pained expression. In fact, I get the feeling he thinks I'm a moron!

Flabbergasted, I finally shut up. This is followed by a long silence. We stare at each other, and then I begin to comprehend. This lawyer is right to think I'm stupid. How could I even imagine for a split second that Alstom would incriminate itself by handing over evidence to the US courts of widespread corruption within the group? Granted, such documents would clearly attest to my minor involvement. But, on the other hand, they would force the company – starting with some of its most senior executives – to admit guilt and acknowledge that the compliance processes implemented by top management were nothing more than smoke and mirrors! Am I foolish enough to believe that Alstom would go along with a line of defence that would implicate them just to help me? Of course they wouldn't. Indeed, what company would voluntarily sacrifice itself, owning up to its criminal liability just to rescue one of its executives? Until then, I hadn't perceived it that way. Whether naive or overconfident, I was loath to imagine the worst. I must now admit that I am in real danger. From here on, I am on my own.

Calmly, I ask Stan once more: 'Have you been able to access the indictment, and have you received any other exhibits from the prosecuting attorney? What is the maximum sentence I could receive?'

'At this stage, it is difficult to give you an answer. Like you, we just read the indictment summary.'

'There is no evidence in it to prove my guilt.'

'I tend to agree with you. No direct evidence. No email from you referring to possible corruption. But the prosecuting attorney will be sending us 1.5 million exhibits.'

'One point five million?'

'Yes, and besides, from what I hear, they have two witnesses who can corroborate your involvement in the conspiracy.'

Finally, to qualify for legal assistance, I end up signing Alstom's document. The conditions imposed on me by Alstom are gross, but I have no choice.

# 9

## Clara

Compared to Stan's cold arrogance, Liz's lack of empathy suddenly appears sweet as honey. She seems to at least have some humanity. As soon as she gets back to her office, she suggests (though it's normally prohibited) setting up a conference call between Clara in Singapore and myself in Wyatt. I'm finally going to be able to hear my wife's voice. This fills me with as much dread as joy. Liz carefully explains to me that she already informed her of my incarceration the day before, when my bail hearing was postponed for forty-eight hours.

'With the time difference, when I called her in Singapore, she had just arrived at her office. Of course, it was a real shock for her. But you told me to inform her as soon as possible.'

'How did she react?'

'At first, she was very afraid. She thought you had had an accident or a heart attack. She called me back several times this morning. I explained to her how to register herself on the list of persons authorized by Wyatt administration to reach you by phone, but it will take a certain time.'

'Can't we speed up the procedure?'

'No, it is just impossible.'

'So when will we be able to speak to each other directly?'

'It depends. Sometimes it takes three days. But with a request from abroad, it may take much longer, perhaps a week, or even two. In the meantime, you will have to go through me. But there is nothing we can do about it; this is the process, and we have to apply it.'

Americans are very proud of their 'process'. I discovered this term when working in Connecticut. Americans love *processes*. In their jobs, they rarely express their creativity. Conversely, they expend a lot of energy and time adhering to their *processes*.

Stan and Liz depart, and four hours later, when I have the right to leave my cell and access one of the prison wall phones, I call Liz on her cell phone. She immediately connects me to Clara.

'Hello! Fred! At last.'

From the tone of her voice, I can sense her exhaustion and fear, despite how affectionate she sounds. Over the last twenty-four hours, she has failed to credit my account, despite her repeated efforts and her multiple attempts to connect to Wyatt's internet platform or her phone calls to the prison. Her bank cards are systematically refused, which drives her to despair. I can easily imagine how her life has been these last few days with her job, which she has to carry out as if everything were normal, my mother who has just arrived in Singapore for two weeks to visit the family, and the four children to whom she has decided not to say anything for now – we mustn't alarm them unnecessarily. After having related the circumstances surrounding my arrest, I try to reassure her as far as I can, even if it means giving her a version that I no longer believe in myself.

'As soon as I was arrested, I was able to speak to Tim Curran, the head of the Power sector in the United States, and also Keith Carr, our legal director. They explained to me that Alstom was in the process of making a deal with the DOJ. They will certainly include my personal case in these negotiations. Tomorrow I will be a free man, and then we will have a better picture thereafter. For now, let's try to keep this matter under wraps.

'It's too late for discretion,' says Clara. 'The *Wall Street Journal* has published an article on your arrest at JFK airport, and *Le Monde* has published a few lines on it. But don't worry, it appears to have gone unnoticed. Your parents have not seen it. And no one from Alstom has contacted me.'

'Good. Well, let's hope it doesn't attract any more attention. I don't want to return to work in a very charged atmosphere! How are the children?'

'They don't suspect anything for now. Gabriella and Raphaella prepared a show for your mum's arrival yesterday; they both dressed up as princesses. Raphaella was the Sleeping Beauty, and Gabriella was Cinderella. Léa and Pierre played guitar. Your mum was thrilled. I decided not to tell her anything. I just told her that your business trip to the United States was going to take a little longer than planned. Well, that's it for now. Frédéric.'

'Yes.'

'I looked up documentation on FCPA cases on the internet. This is really very serious stuff. I found that this law allows Americans to arrest any employee of a company, anywhere on the planet, at any time and incarcerate them for a very long time.'

'Aren't you exaggerating a little?'

'No Fred. I don't want to worry you, but the US authorities consider that the slightest link with the United States, stock market listing, trade in dollars, use of an American mailbox, authorizes them to take action. It sounds a little crazy, but it happens a lot. Moreover, Alstom is far from being the first French company to be prosecuted for corruption. I checked: it has already happened to Total, Alcatel, or Technip. And that's not to mention the dozens of other major European companies that have also been indicted.'

'Have any executives been arrested, like me?'

'Yes, apparently in the Alcatel case. And then, in an investigation opened against Siemens, the FBI even launched international arrest warrants against German senior managers. I get the impression that ...'

I sense that Clara is very concerned by what she has read. At the same time, she is reluctant to tell me all. She probably doesn't want to give me any more cause for alarm.

'What impression do you get?'

'You know the Americans; when their interests are at stake, they don't cut any slack. And ...'

'Go ahead. I'd rather you tell me the truth.'

'Well, even if they release you on bail, the question I have is, will they force you to stay in the United States?'

I don't know what to reply to her. How do you announce to your wife that her life is about to be turned upside down? I have no idea how this is going to end. As someone who likes to master all the ins and outs of a question before making a decision, I find myself in limbo. But Clara is ahead of me.

'Fred, if necessary, we could come to you in the United States in no time. We've spent our lives on the move, so it's just one more journey ... Don't worry, if you need me, I'll come with the children. You know very well that they are used to this nomadic lifestyle. Please do not worry. I am with you.'

I am moved by her sheer resolve. Clara has the ingenuity to project herself into the future at breakneck speed. But for the moment, the future is limited to my second bail hearing scheduled for tomorrow. Until then, we are all in limbo. Subject to the whims of the American judiciary and their 'processes'.

# IO

## The second bail hearing

This time, they did not forget me. At precisely 4 a.m., two guards burst into my cell to awaken me. I am forced to undergo a new body search, before being chained from head to toe as I was during my initial transfer to Wyatt. Then they put me in an armoured truck, heading for the New Haven Court of Justice three hours away.

A few minutes before the start of the hearing, I am allowed to talk to my defence counsel, Liz and Stan.

They look drawn. They tell me they have just been talking to Assistant US Attorney David Novick.

'He won't back down,' Stan tells me. 'The amount of the bail we are prepared to put up is of little importance to him; he wants you to remain behind bars. I think your company's lack of cooperation has stuck in his throat. He feels that Alstom has been taking him for a fool for several years now.'

I continue to slide down, and down, as if I'm never going to touch the bottom. What I discover at this point, reading between the lines of Stan's remarks, is that the Department of Justice began its first investigations over three years ago. The investigation has therefore been ongoing since the end of 2009. Keith Carr has been very careful not to inform me of this chronology. All of a sudden I realize why the DOJ is pursuing me so relentlessly. They need a fall guy, someone to take the blame for Alstom's doublespeak.

As soon as the Americans began their investigation, they alerted my company and asked it to cooperate. The DOJ has a process whereby each company under investigation receives an official offer to cooperate. Such companies are then given the opportunity to sign a DPA (Deferred Prosecution Agreement). To do this, the company must agree to incriminate itself by disclosing the entirety of its practices, and if necessary, by denouncing its own staff. It must also commit to setting up an internal anti-corruption mechanism and accept the presence of a monitor, i.e. a controller who reports to the DOJ for a period of three years. Provided all these conditions are satisfied, the court ratifies the deal made with the company, which generally culminates in a monetary penalty. As a rule, senior managers are not subsequently arrested (though, in theory, this type of agreement does not protect individuals from prosecution). It is precisely in this way that prior to Alstom, two other French companies, Total and Technip, agreed to pay fines of \$398 million and \$338 million in 2013 and 2010 respectively. However, Alstom, or rather Patrick Kron, as I would later discover, decided to try to outwit the DOJ by letting the DOJ believe that the company would co-operate, while in reality doing the opposite. When the Department of Justice realized that it had been fooled, its state attorneys went mad, and decided to change their strategy. They went on the offensive.

Therein lies the 'real' reason for my bolt-from-the-blue arrest. The DOJ has now decided to show Alstom who holds the most cards, to force it to plead guilty. I have been stitched up, victim of Patrick Kron's game plan and reduced to dummy-like status in the hands of the American judiciary. In a few moments, I will have confirmation from US Attorney David Novick himself. When Joan Margolis, the judge chairing the court hearing, gives him the floor, Novick starkly reveals the backdrop of the battle he has been waging against Alstom.

'After pledging to cooperate, this company has repeatedly betrayed the trust of the Department of Justice. In reality, Alstom, who was supposed to assist us in our investigation, has acted in a very slow and piecemeal way. Its attitude is nothing less than ambiguous. I would also like to mention that the OECD has recently pointed to the attitude of France, which has failed to take measures against Alstom, whereas this multinational has been the subject for several years now of allegations of corruption in numerous countries, including Switzerland, the United Kingdom, Italy and the United States.'

The US attorney is unrelenting in his will to prosecute.

'We have all the evidence. Documents reveal the discussions between the co-conspirators to bribe Indonesian government officials. We also have bank attestations, and we have witnesses who will come forward and testify.'

I wonder how the American judiciary could have obtained these documents. During our discussion, which preceded the opening of the proceedings, Stan and Liz astounded me by revealing what lay behind this battle. At first, I did not believe them. The manipulation appeared too blatant and resembled a movie script. Well, I was wrong. This is for real. To obtain evidence against Alstom, the DOJ had, inter alia, used the services of an 'executive turned informant', a snitch placed at the very core of the company who cooperated fully with the investigators. For several years he even wore a microphone hidden under his jacket, which allowed him to record numerous discussions with his colleagues. The FBI used him as a mole in the firm. How could an executive accept this role at nearly sixty-five years of age? What pressure did the FBI and DOJ officials exert on him to compel him to become a 'traitor'?

Did they threaten him with a long jail sentence? I barely have time to delve deeper, because attorney Novick is now addressing my case. He doesn't go in for half measures. According to him, I am one of the architects of one of the most astounding, brazen and extensive corruption scandals he has ever witnessed throughout his career.

'Frédéric Pierucci is a very senior executive at Alstom. Over the past few years, his leadership has consistently entrusted him with major responsibilities. His firm is now offering to post bail of $1.5 million to have him released. But his firm is implicated in this case. It is a "co-conspirator" even though it has not formally been charged. Therefore, the question I have is: can a "co-conspirator" stand surety for a bail payment? And besides, we still don't know exactly what turn this file is going to take. What will happen if, at some point, the interests of Alstom and Frédéric Pierucci diverge? What will happen if this senior executive decides to plead guilty? How will Alstom react then? Who will then be liable for the bail? Don't forget that the use of electronic tagging is not an adequate insurance. Frédéric Pierucci could disconnect it at any time and flee. As for the guards that his company agrees to pay for his surveillance, what will happen if his firm suddenly decides to stop paying them?

'Two years ago, the Department of Justice authorized the use of guards to monitor the residence of another French citizen, Mr Dominique Strauss-Kahn. But this case is very different. In the D.S.K. case, when the judge agreed to this arrangement it undermined the prosecution. The main witness for the charge against Dominique Strauss-Kahn was discredited. But here, on the contrary, we have a very solid and well-founded case. And as you well know, Your Honour, France does not extradite its citizens. Frédéric Pierucci represents a flight risk! He is fully aware that he faces a very stiff sentence in the United States . . . life imprisonment.'

Life imprisonment! I immediately turn to look at my lawyers. Stan Twardy averts his eyes. Liz Latif is also afraid to catch my gaze. She is more at ease focusing on her notes. I'm facing life imprisonment. A life sentence! I am forty-five years old and may have to spend the next thirty or forty years behind bars. I've only been in Wyatt jail for five days now and am already at the end of my tether. I don't even know if I can handle this nightmare for the next few hours, let alone for the rest of my life. And on what grounds? Because, ten years ago, I was one of the Alstom executives who endorsed, at middle manager level, the selection of a consultant, whom I didn't know from Adam or Eve, who allegedly paid bribes to help the group win a contract. I have neither stolen, nor injured, nor killed anyone, nor received the slightest kickback. I have not benefited from any personal gain. To top it all, everything was carried out in strict conformity with Alstom's internal processes. Life imprisonment! It's just absurd. This is blackmail. This attorney is trying to intimidate me. He's messing with my head.

Yet, when he sits down, David E. Novick, unlike my two lawyers, stares me straight in the eye; he isn't bluffing, he looks deadly serious. What if it is true? What if I really do risk spending the rest of my life in prison?

I'm so shocked that I barely even hear Liz Latif asking for my release. Her voice sounds faint and blurry to me, as she attempts to point out a procedural error in the indictment. She claims that the charges against me are time-barred because they were allegedly committed between 2003 and 2004, whereas my indictment is dated November 2012. However, the prescription period is five years for FCPA violations.

So why is she bothering with this legal pitch? Why doesn't she just tell the truth? Is it so hard to explain to the court that I am being subjected to unjustifiable differential treatment? David Rothschild, who has already pleaded guilty, was not incarcerated. And he only had to put up $50,000 bail, which is disproportionate compared with the amount proposed for my release. Is it lawful that attorney Novick should use me as leverage in the power struggle he has mounted against Kron? He makes no secret of it: he is holding me hostage. I am a pawn in the game of chess he is playing with Alstom. Is that justice?

Judge Margolis takes recess for a few moments to deliberate. When she returns, I immediately realize that the case has been decided.

'This is an unusual case for our court,' she says. 'Usually we deal with requests from very modest families. The sureties do not exceed one and a half thousand dollars and these often represent life savings. In this case, the defence has proposed bail of more than \$1 million. However, I deem this to be insufficient. I would like to see, in addition to the one and a half million dollars pledged by Alstom and the four hundred thousand dollars proposed by the defendant, an American citizen commit to putting his own home up for bond. If you come back to me with this commitment, then I will agree to reconsider your release on bail.'

In short, Joan Margolis distrusts Alstom and questions my credibility. To convince her to change her mind, she requires a pledge by an American citizen. By contrast, attorney Novick had no difficulty in persuading her. He strides out of the courtroom, stiff as a post, looking like the cat with the cream.

I am totally devastated and having a hard time containing my anger against my defence lawyers. The astronomical amount proposed by Alstom as surety was a strategic miscalculation. This solution has backfired on me. I placed my trust in them; how could I have done otherwise? And I made a monumental error. How could a lawyer as experienced as Stan, himself a former United States Attorney for Connecticut, not see the major threat of a possible conflict of interest with Alstom, and not anticipate the reactions of Novick and the judge? I start to have serious doubts about him. Who is he really working for? At the end of the hearing, I also discovered that an attorney from Patton Boggs (representing Alstom) was present in the chamber as an onlooker. He was watching my every move.

Alstom can rest assured: I didn't say a word. Oh yes, I have received the message loud and clear: I am under surveillance, trapped between Alstom and the DOJ, and represented by a lawyer I didn't even choose.

# II

## 125 years in prison

I thought I would never see these walls again ... I'm back in my cell in Wyatt. And I could be there for a long time, even several weeks; however long it takes to be able to make a new bail application: the third.

In prison, while every hour seems like an eternity, I still have no news from Alstom. Stan told me that Keith Carr, the general counsel, had been to Washington to negotiate with the Department of Justice. He arrived just twenty-four hours after my arrest, but it appears the FBI did not question him, which continues to puzzle me, as this experienced lawyer has held a strategic position in the company for the past ten years.

He knows its nuts and bolts. In 2004, he was deputy general counsel for the Power division. A year later, he was appointed Power Sector's general counsel before being promoted in 2011 to group general counsel. He is therefore well acquainted with all the company's business 'practices'. He knows better than anyone how Alstom recruits and remunerates its 'consultants'. Why didn't the investigators arrest him? They would have certainly learned more from him than from me. Why did they target me? This is what I can't understand.

I had hoped that Keith Carr would also use the opportunity of being in the United States to visit me in Wyatt, yet no sight nor sound from him. No response from the other company officers either. I know there is no sentiment in the business world, but this does not stop me from feeling profound disgust. Overnight, I have become the black sheep of the group. A shady individual, not to be associated with, to be avoided like the plague. I was expecting a bit more peer support from general management, with whom I have worked for twenty years.

Stan and Liz come back to see me midweek. We are to search for people in my circle of friends and professional contacts in the United States who are prepared to put their houses up for bail bond so that I can get out of jail.

'As you suggested to me,' says Stan, 'I have asked Tim Curran, the head of Alstom Power in the United States, and I have also made the request to Elias Gedeon, vice-president of sales. They both said no. They both said that it is up to Alstom, and not them, to ensure that the judge is satisfied.'

'Frankly, I understand them,' I remark to Stan. 'I wouldn't have taken that risk myself.'

'Do you have friends or relatives in the United States?'

'Very few. We left the country seven years ago and we have no family here. I have kept in touch with some people, but we're not that close. Clara is nevertheless contacting everyone. Our greatest hope is one of her closest friends, Linda. We are waiting for their replies, including hers. What if we proposed to put up our home in France as bond?'

'No, the judge will refuse. The American judiciary has had great difficulty in the past in seizing property in your country.'

I'm back in the slide. An endless tunnel with smooth walls. Nothing to grip. As soon as I see a solution, it slips away. I already know what Stan is going to announce. To put it mildly, he doesn't beat about the bush.

'For the time being, you will remain in prison,' he summarizes. 'This morning our office received a first proposal for a trial date: 26 June 2013. In two months.'

I try to hold onto the slippery walls of the endless tunnel.

'But if Alstom includes me in its deal with the DOJ, that should change the game, right?'

'I'm afraid not,' says Stan. 'These are two different prosecution cases. The DOJ can indict a legal entity and make a deal with it, but that doesn't prevent them from prosecuting you, as an individual.'

'That, I understand. But they can still include me in their deal.'

'In theory, yes, but it is more difficult after an indictment, which is the case here. I also doubt that they will do so as their lawyers will try to persuade them otherwise, to minimize the fine and, above all, to protect the other executives who haven't been charged.'

'And can I negotiate too?'

'Yes, you can plead guilty.'

'I meant negotiating a fine in exchange for my release.'

'No, you can plead guilty and it will be up to the judge to decide whether or not you are given a prison sentence.'

'Therefore, if I plead guilty as did David Rothschild, I risk five years in jail …'

The end of the tunnel. A long tunnel. But at least I can see the end. In vain, Stan tells me that I am not even allowed this poor fate.

'Unfortunately,' he explains to me, 'your situation is more complex than that of Rothschild. He was approached by the FBI first and he agreed to cooperate immediately; he therefore secured the best negotiating terms. Undermining your own bargaining position. You are in second place in the race to satisfy the prosecution and therefore less able to help them in their investigations, and you turned down the deal that Novick proposed to you.'

Deal? Market? Negotiation! Since our briefing began, Stan and Liz only ever talk to me about negotiating, and never about the case being judged on facts and evidence. This is like haggling at a horse traders' fair, except I am the horse they are trading. Okay, if that's what they want, full steam ahead. Let's be pragmatic, as they say here. To hell with justice and truth. Let's make a deal. Let's put aside sentiment and the intolerable feeling of knowing that you are the fall guy. Let's use their language. I take a deep breath.

'Okay, let's start over, if you don't mind, Stan. What kind of sentence am I facing? Novick is threatening me with life imprisonment. I guess this is intimidation; it can't be for real?'

'Uh, well,' says Stan, 'in theory, he's not far from the truth … You are being prosecuted on ten counts. The first is a charge of "conspiracy of FCPA". In a nutshell, you're accused of conspiring with other executives to bribe an Indonesian government official who was a member of the Energy Commission at the parliament of Jakarta, in order to obtain the Tarahan contract. This offence is punishable by five years' imprisonment. However, as prosecutors have the evidence to prove four successive payments of money to a relative of this parliamentarian, you are therefore prosecuted for the first offence of conspiracy, to which are added the four payments of money, which are each time considered as an additional offence. You therefore risk being sentenced five times to five years in prison, totalling twenty-five years. To this, we must add the second major offence that you have been charged with: that of "Conspiracy of Money Laundering". This money-laundering offence is punishable by twenty years' imprisonment. Here also, based on the verified payments made, this penalty must be multiplied by five. You are therefore facing a hundred years in prison for money laundering and twenty-five years for corruption. Thus, theoretically speaking, we arrive at a total of one hundred and twenty-five years.'

It's no longer a tunnel. It's a chasm. I want to let out a nervous laugh. But I desperately try to reason.

'Wait, Stan, but this is totally insane. We are dealing with a single consultant contract! How can an attorney prosecute ten counts of FCPA and money laundering from the same offence?'

'That's how our judicial system works, Mr Pierucci. Our definition of money laundering is not the same as yours in Europe. Here, from the moment that a monetary transaction is illegal, the DOJ considers that money laundering has also occurred.'

'But this is stupefying! I need you to give me a lot more information on FCPA and its case law.'

Stan stiffens up as though it were a personal attack.

'I don't think this is the time to discuss it,' he snaps. 'Our priority is negotiating the best deal for you, isn't it?'

We're back to that. Negotiate, strike a deal. I knew it, I had read and heard about it: America's judiciary is a marketplace. But you have to have experienced it to understand what it means. I have already started to wonder whether lawyers, judges, attorneys are all connected by the same 'commercial' ties. I am seeing everyone as a trader in the same marketplace, with no idea whom I can trust. One hundred and twenty-five years in jail! Why this doomsday scenario? Is he also trying to break me? I toss another question at him.

'To negotiate, as you say, I first need to know a bit more about the charges that the prosecuting attorney has brought against me. What evidence has he presented? How am I personally involved? It's been over a week since I was thrown in jail, and you still haven't told me anything concrete yet.'

This time it's Liz who cuts to the chase.

'The indictment that details the charges against you is seventy-two pages long ... Let's start reading it and get down to work.'

# 12

## The indictment

The indictment against me is entitled: 'The United States of America versus Frédéric Pierucci'.

I feel gutted at the sheer title. The document consists of ninety-one paragraphs, plus about forty pages of appendices, mainly copies of emails I received or wrote when I was based in Windsor, Connecticut. The prosecution refers to at least twenty of them. The first of them dates back to over eleven years ago, i.e. February 2002. In the early 2000s, I remember that there was tremendous pressure on managers. We had just one mission: to save Alstom from bankruptcy. Top management therefore urged us to deploy our absolute maximum efforts to win as many bids as possible. 'Results over effort' became the official catchword. It was against this backdrop that we approached the Tarahan project in Indonesia, which involved the construction of two 100-megawatt boilers, for a total sum of $118 million. For Alstom, this represented a relatively small-scale contract. But it was one of the few we were able to secure in this troubled period. It quickly became highly strategic, labelled 'priority' by top management.

But one of the concerns was that this business was located in Indonesia. At the time, Indonesia was one of the most corrupt countries in the world, although the situation had improved slightly following the departure of the dictator Suharto and his family in 1998. Under this dictatorship, fully supported politically and economically by the United States, it was not uncommon for companies to shell out 15 per cent or even 20 per cent of a contract's value in commission to middlemen who were close to the Suharto family. Be that as it may, with or without Suharto it was common knowledge that no deal could be negotiated in Jakarta without paying bribes. On the other hand, we knew that we had a real chance of winning. In fact, when the dictator was in power, most of the contracts were awarded to American or Japanese corporations. As for the power business, an American duopoly swept the market, comprised of Babcock Wilcox and Combustion Engineering, ABB's American subsidiary, which had just been acquired by Alstom.

As a result, we were well placed to compete in this bid for the Tarahan project. In addition, PLN (the state-owned and state-controlled electricity company in Indonesia) had chosen a technology that was well suited to our product range: using 'circulating fluidized bed', a so-called 'clean coal' technique used to burn difficult-to-combust coal by extracting a very high percentage of pollutants. Alstom was one of the two global leaders in this high-tech sector with another American rival, Foster Wheeler. In other words, things were looking pretty good. And this is how events unfolded.

One day in August 2002, I was contacted by the regional sales manager in the Windsor team, David Rothschild. He asked me to endorse the hiring of a consultant to help us obtain the contract. Indonesia being one of their major long-established markets, I presumed the Windsor team knew how to manoeuvre in this hotspot. On 28 August, I replied to him by email. An email that I find transcribed verbatim in the indictment issued by the prosecution. Exhibit 43 of the indictment: 'Go ahead and send me the key data so I can officially approve it.'

I remember it very well. And I recall, immediately after sending this message, that I was gripped by doubt and summoned Rothschild to my office to obtain more information on this consultant we were about to hire. Rothschild then tells me, as if it were the most natural thing in the world, that the person in question is the son of a certain Emir Moeis, an Indonesian parliamentarian in charge of the Energy Commission. At that time, the FCPA mechanism to combat the corruption of foreign public officials, which had existed in the United States since 1977, did not seem to overly worry my new American colleagues. And why should it, given that until 2002 this law had been very poorly enforced in the United States (less than once a year) and that neither Combustion Engineering nor any of their American competitors in the power generation sector had been troubled by it in twenty-five years.

Personally, even if, at that time, I had little knowledge of these legal aspects, the payment of a commission to the son of a parliamentarian nonetheless seemed like crude abuse to me. I therefore immediately ordered Rothschild to promptly abort the recruitment of this consultant. I know that I was walking on eggshells, as this consultant was selected by Reza Moenaf, a key figure who headed the Alstom boiler unit in Jakarta and piloted Alstom's international network in Indonesia. Rothschild heeded my recommendations and executed my orders.

The prosecutors also cited the email he sent in the wake of my recommendation (Exhibit 44 of the indictment), 'Do not finalize! I have discussed it with Pierucci. We have concerns about the politician.'

I suspected that halting the process in this manner would make me a few enemies, but at this point I had no idea just how many.

I found out later on that my refusal probably blocked the payment of kickbacks to certain individuals who would subsequently make me pay dearly in return.

A few days later, at the beginning of September 2002, Rothschild announced to me that Moenaf had found a new consultant, a certain Pirooz Sharafi, an Iranian-born American living in Washington but who spent half his time in Indonesia where he had been running a business for many years. His address book was as thick as a phone directory. I also learned that Sharafi had already successfully worked for ABB as a consultant on other Indonesian contracts, or so Rothschild assured me. This new consultant was depicted as a high-profile lobbyist who would never resort to bribery. I have some major doubts of course, but it is tricky for me to once again oppose the ex-ABB in Windsor. Either way, it is up to the compliance teams in Paris to carry out investigations into the integrity of such consultants. My function is simply to integrate the cost of these intermediaries into the sale price. The rest is out of my hands. I was aware of some of what happened within the company at the time, but reading the indictment against Alstom filled in some of the gaps in my knowledge. This is what seems to have happened with regard to the appointment of consultants on this contract.

A few months later, our future consultant, Pirooz Sharafi, was received at Alstom's headquarters in Paris accompanied by Emir Moeis, the parliamentarian in charge of energy in Indonesia. On this occasion, he was also introduced to Lawrence Hoskins, senior vice-president International Network for the Asia region for Alstom (two hierarchical levels down from Patrick Kron). Thereafter, Sharafi also met with the heads of the compliance department. There was even talk of using his services for a second contract in Indonesia. At the end of his Parisian visit, the Compliance department and the International Network department approved of his hiring as a consultant for the Tarahan contract. Our management seemed highly satisfied with the assurances provided by this consultant. What more could you ask for?

Sharafi's mission was straightforward, and involved organizing interviews with clients, politicians, financiers and consulting engineers to extol the merits of our offer. In short, the usual role of a lobbyist. His remuneration was fixed at 3 per cent of the overall value of the sale. Again, this percentage is quite typical for this type of service.

In the months that followed, Sharafi set to work. On this contract, we were partners in a fifty-fifty consortium with a Japanese counterpart, Marubeni (which also validated choosing Sharafi as consultant), and in competition with a huge American corporation. At first, everything seemed to be going well. We learned that our offer was the most competitive and the most technically sound. In other words, we were the best placed and should not lose the deal. However, we knew that outpacing an American corporation is a delicate issue, especially in Indonesia, a country that is located in the United States' zone of influence. So together with Sharafi and Moenaf, I visited the US Embassy in Jakarta to plead our case and try to appease the American giant's concerns.

I guess I didn't come up with the right arguments, because in the summer of 2003, there was bad news. Against all odds, the situation boomeranged in favour of our American rival. Clearly, its consultants 'won over' the PLN evaluation team, and Moenaf was informed at the time that they did so by paying or promising to pay kickbacks to key individuals. We were in danger of losing it. Our Japanese partner arrived at the same conclusion. Marubeni's CEO then contacted Patrick Kron to convey his concerns about Tarahan. The prospect of failure in this negotiation sent shock waves through the Paris headquarters. I immediately received orders from my boss to head straight for Indonesia to get the deal back on track. Senior management also told Lawrence Hoskins, the boss of Alstom International Network Asia, to go in person. I was under immense pressure.

In turn, I urged my staff to do everything they could to reverse the situation. The prevailing state of panic is evident in one of my countless emails that the prosecution is brandishing today (Exhibit 55 to the indictment).

For instance, on 16 September 2003, I wrote to William Pomponi, who replaced David Rothschild as regional sales manager on this case, and who was located in Jakarta: 'When we spoke on Friday you told me that everything was under control. And now I hear we are in second place! Give me a plan tomorrow to get it back. WE CANNOT LOSE THIS PROJECT!'

A crisis meeting was held at the end of September 2003 at the Hotel Borobudur in Jakarta. Marubeni, our Japanese consortium partner, had come to the same conclusion as we had that our American rival had promised to bribe several members of the evaluation committee and PLN management. The Alstom International Network teams agreed with the Japanese to hire a new agent, known as Azmin. I didn't know this guy Azmin, and I never met him, but they had already worked with him on another Indonesian contract: 'Muara Tawar 2'.

My job was to ensure that the use of this second consultant would not be a financial drain on us. Ultimately, it was decided to reduce the rate of the first consultant, Sharafi, so that the second consultant could receive a larger payment. Sharafi would only receive 1 per cent commission, i.e. about $600,000, while the remaining 2 per cent would now go to Azmin. Easier said than done, but time was running out. Lawrence Hoskins promptly referred the matter to the Paris headquarters, and top management gave him the green light within twenty-four hours.

In September 2003, Azmin stepped into play. He was a resounding success, and in 2004 we were finally able to clinch the deal. However, the use of two consultants required Alstom's management to set up two parallel financing circuits. The American unit in Windsor paid Pirooz Sharafi's remuneration, and Alstom's Swiss subsidiary (Alstom Prom) was responsible for Azmin's remuneration. Alstom admitted later in its plea agreement with the DOJ that its own Alstom Prom's compliance department assisted consultants to create 'fake proof of services' to justify their consultancy fees. This is, most probably, what happened for Azmin to be able to get paid.

Just before the final signing of the Tarahan contract on 24 May 2004, Eddie Widiono, PLN's boss, was given red-carpet treatment when he was received at Alstom's headquarters. To welcome him, Patrick Kron surrounded himself with his entire staff. As for me, I attended the post-reception lunch with the group's chain of command. They were all perfectly aware of the conditions for obtaining this future contract, which was ultimately signed on 26 June 2004.

After that, I practically lost track of this file. All I know is that Sharafi (the first consultant) only received the last part of his payment very late, in 2009. Then, mid-2006, I was posted back to France.

Today, the prosecuting attorneys are accusing me of having been one of the 'co-conspirators' in this corruption case. But I repeat vehemently to my defence counsel, Stan Twardy and Liz Latif: 'I received no personal gain whatsoever from this contract and didn't touch a single dollar in kickbacks. In the indictment, the prosecutors are very explicit on this point. If they had any doubt, they would have written it. I was just doing my job, that's all! As my hierarchy requested me to do at the time! So what am I doing in prison? And why me, and not the others?'

'That is correct,' replied Stan, 'but you were well aware that what you were doing was illegal or you at least knew that you were in a grey zone, didn't you?'

'Yes, I can't deny we were in a grey zone. But I do not know of anyone holding a similar function to mine in a French multinational in the early 2000s who would have denounced such practices.'

I am enraged by the look of resignation on the faces of my American lawyers. Obviously, they don't want to or cannot understand my reasoning.

And then Stan Twardy releases information that ruins what little hope I had left.

'There is one last thing you need to know. In addition to emails and conversations, the attorneys also have witness statements at their disposal.'

'I know, Stan, David Rothschild's testimony . . .'

'Not just his. Pirooz Sharafi has also been doing a lot of talking. He was most probably the first one to rat on you. The FBI interrogated him about a tax evasion case. To avoid a lengthy sentence, he made a deal with the investigators. In exchange for full immunity, he traded in the Tarahan case and thus blew the whistle on you. You and others.'

# 13

## Jail becomes a way of life

For the first time since the beginning of my incarceration, I woke up a little less tired than the previous mornings. I'd finally managed to sleep. My two cell mates, Cho the forger and Mason the dealer, even claimed that I snored all night!

'You see, Frenchie,' they joke, 'we did warn you: you eventually get used to anything; even jail.'

At the same time, and as if to prove otherwise, I reply: 'No, you never get used to jail, especially not Wyatt.'

A member of staff knocks on the door. He orders us to get out right now into the corridor. Full search of the establishment.

A few seconds later, a dozen men in black, wearing helmets and armed like the SWAT teams you see on TV, burst into our cell. They are accompanied by the warden, who is escorted by two prison staff. The raid begins. They go through everything, mattresses, covers, sheets, pillow-cases. Every nook and cranny is scrutinized, and everything is turned over. Then, one by one, we have to undergo a strip search in the communal shower room, before being able to return to our cell. We are then summoned to individual interviews with a counsellor.

'Tell me, Mr Pierucci,' she asks me very seriously. 'I know you've only been with us a short time, but have you noticed anything strange?'

I find it difficult to contain a nervous laugh. Is she for real? In Wyatt EVERYTHING seems strange to me! While the screws are turning the prison upside down, I don't feel this is the most appropriate time to be sharing my grievances with her. I prefer to remain silent. She reels off her questions mechanically.

'Have you been the victim of any violence? Have you witnessed drug or medication trafficking? Have you heard any rumours? Have you witnessed inmates assaulting each other?'

Does she take me for a fool? Does she really believe that Frenchie, the white-collar offender living in the midst of hardened criminals, is about to shatter the prison's code of silence?

Does she want me dead or what? This time I really haven't seen anything. But my silence doesn't trouble her at all. She did her job. She asked her questions. She dismisses me with no qualms. Direction: the toilets. A guard hands us each a small vial in which we must urinate, in front of him. It's a test to check that we're not secretly using narcotics. Negative! Back in our cells. Clunk click. The cell door closes.

It will remain so all day long. Too bad if we can't use our wash basins, or toilets. The water was cut off during the raid of the detention facility due to concerns that some prisoners might get rid of narcotics or other items in the drainage system. In the evening, we learn that three inmates were sent to the hole after the raid.

The next day, Cho leaves us. He is being transferred to California to serve the rest of his sentence. It's not very good news. The young Dominican who now occupies his bunk remains lifeless for hours on end, on his mattress, his eyes wide open and with a glazed look. At times, he screams out delirious things. His brain is obviously fried from too much crack use. I can no longer breathe in the confines of this minuscule cell.

Fortunately, Clara has managed to credit my commissary account from Singapore, and we can now talk on the phone. I will also be able to make my first purchase at the commissary: a toothbrush, toothpaste, a shaver, shaving foam, cotton swabs, a few changes of clothes and underwear.

The only positive thing for me about the days passing during my detention is that my prison quarantine period is over. I can now leave my cell, go to the common room, access phones, meet other prisoners, discover the microcosm of Wyatt, with its litany of horrors and also its humanitarian side.

This is Chris. He's a true gangster. His 'track record' is impressive: about twenty bank robberies and as many convictions. At age fifty-seven, he has already spent twenty-six years behind bars. He has two children: a twenty-six-year-old son he has never seen and a daughter he conceived in a prison visitation room. He also has three grandchildren but doesn't even know what they look like. On the other hand, he knows the American prison system by heart.

He has already served time in a dozen federal penitentiaries and is obsessed with lawyers.

'Frenchie,' he repeatedly tells me, 'never trust your lawyer. Most of them are working undercover for the government. And above all, never admit anything to your counsel; otherwise, he'll force you to make a deal on his terms, and if you don't, he'll rat everything to the district attorney. And beware of other inmates too. There are a lot of snitchers. As soon as they learn something, they rush to repeat it, to try to get a lighter sentence.'

Chris sees conspiracies everywhere, all the time. He is convinced that as a former Attorney General, Stan is the wrong lawyer for me, as he is too well connected to the DOJ, and advises me to replace my defence counsel with his.

'You won't find better. He even represented the Hell's Angels!'

And worst of all, I get drawn into his paranoia and spend the whole day considering whether to heed his advice, before coming back to my senses.

However, Chris's warnings continue to echo in the back of my mind as my story unfolds.

In what hell on earth have I landed? How could my life have turned upside down in just a few days? I feel like the whole world has abandoned me, except my family. Although, to my great surprise, I did receive a very special visit yesterday afternoon.

'Pierucci! Lawyer visit', shouts a member of staff. Meetings with lawyers, or prison administrator representatives, take place in a separate room in Wyatt.

These are 'contact visits', unlike the others, which are known as 'non-contact visits' where a glass partition separates you from the person you are speaking to. After having to pass through thirteen armoured security doors and undergoing another full strip search, I entered a room where a young woman was waiting for me.

'I have been sent by the French consul in Boston. The consul wanted to come in person to meet you, but at the last moment he was detained.'

Very slim, in her forties and elegant, Mrs L. hardly seems at ease in her role of prison visitor. Visibly shaken by the setting and seemingly anxious by nature, she forgets the essentials. Instead of enquiring about my situation and the support she can give me, she relates her life story: her most recent stay in Indonesia, the great romance she encountered there, her son's ordeals, and even what she ate in a fancy restaurant.

I listen to her, dumbfounded, not knowing whether to get angry or let her continue to offload her inappropriate small talk without interrupting her. But why is she visiting me? I finally get it at the very end of our interview. As the consul's envoy is about to get up and go, she suddenly becomes very professional again.

'One last thing, Mr Pierucci. Don't consider requesting your transfer to a French prison. The Americans will not let you go until they have delivered their own verdict. They feel that the French are far too lax in prosecuting corruption matters.'

So that was the purpose of her visit. Message received loud and clear. I'm on my own. My own country won't make a stand.

A few weeks later, however, Mrs L. was replaced by Jérôme Henry, Deputy Consul in Boston. Unlike his colleague, he proves to be extremely helpful. Feet on the ground, pragmatic and compassionate, this diplomat was to visit me several times during my detention and maintained close contact with my family. Throughout the whole period, he was one of the very rare and invaluable sources of support I received, notwithstanding his very limited room to manoeuvre.

For now, I return to my cell, following this surreal encounter. Through the one and only narrow 'window' in the shape of a loophole, I see a first fence, then a few metres further on, a row of barbed wire, then another fence, and further on a hill, on top of which I perceive a marmot. I observe it obsessively. Whether it is due to the presence of this animal or the blue sky engulfing Wyatt that morning, they allow me out for the first time into the yard. It is 15 degrees, the weather is cool, and the sky is immense. It is a spring morning on the east coast of the United States, just the way I like it. I am by myself and try a few basketball dunks. I almost feel like I'm free. Countless questions jostle in my head. What is my father doing right now? Has Raphaella, my youngest daughter, finally found the right medication to prevent her hair loss? What did Mum do during her stay in Singapore? And Clara, how is she coping with the four children? And the tax return: will she be able to fill it out correctly? And ...

# 14

## Family support

Clara is coping well, extremely well. In Singapore, she took matters into her own hands. This past week she has moved heaven and earth to find an American citizen willing to put his/her house up for bail bond.

Without this guarantee, no judge is willing to grant me release on bail. My wife also approached Alstom. She discussed my situation at length with Mathias Schweinfest, legal director of the Thermal Power division, headed by Andreas Lusch, my boss for seven years. Mathias Schweinfest reports to Keith Carr, group general counsel.

'It was a shock,' she tells me over the phone. 'He started off being nice. He explained to me that the company would support you, but then when we got into the details, I realized that it was all hot air ...'

'What do you mean? What did he say exactly?'

'Well, Alstom is not willing to help us find someone who will put up their house for bail bond. The American judiciary is against it. The prosecution believe that it would be a subterfuge, a form of disguised assistance.'

I knew that already ...

'But the worst is yet to come. If we manage to find someone ourselves, i.e. an American citizen who is prepared to put up bond for you and you are released, then you will no longer be able to work for Alstom!'

'What, how come? That's impossible, they can't do that to me!'

'Yes, they can Fred. Mathias Schweinfest was very clear on this point. Since Alstom is also under investigation by the DOJ, you will most certainly be prohibited from meeting with your colleagues in Windsor, at least throughout the entire investigation. I'm not even allowed to contact them any more. They have banned me from speaking to any of Alstom's management!'

I feel like I've been hit on the head with a club. My future plans are falling apart. Until that moment I had come to terms with the idea that, after my release, even if I had to stay in the United States awaiting trial, I would have the opportunity to carry on working from the group's American headquarters in Windsor, which houses the American boiler business unit. Now, after what Clara just told me, everything is in disarray. It doesn't merely affect me. Without Alstom's help, how will my family cope?

Clara reads my mind.

'At least until you plead guilty, they'll continue to pay you. But at the rate things are going, you may have to bring Alstom before the employment tribunal at some point. It's disgraceful that they have cut you loose like this! I'm a bit disoriented, but don't worry: I'm in contact with all your family, especially your sister, who is of great help to me. I also found some interesting information on the FCPA online, including a report prepared by a law firm that specializes in this field. I'll send it to you.'

I know how precious my sister Juliette's assistance and support must be. She has a solid background in law and as she tells me in a letter, she has already analysed the indictment against me (the document was almost wholly available on the US Department of Justice website the day after my arrest).

'Dear Fred,' writes Juliet in her letter, 'when Clara informed me of your arrest by the FBI, I was so shocked that, trembling, I had to sit down immediately on the steps bordering the market square. I could sense Clara was on the brink of tears. As soon as I got home, I saw, just by typing your name in Google, that you had been charged: *Indictment F. Pierucci v United States*. In other words, that you have been indicted by the American government. To my great surprise, by clicking on the Department of Justice online document, the attachment opened, and I was able to consult a text of nearly seventy pages!

'For us French, it is simply stupefying that the public have access to these charges before you have even been tried! I read the whole file in detail. How have they been able to detain you? Based on what evidence? I am truly outraged by the way their justice system works. And even if all this were true, in France and Europe if acts of corruption are proven, it is first and foremost the companies themselves and not their employees that are targeted, unless managers have acted in their name or have made personal gain, which is obviously not your case. Fred, don't be discouraged, hang on in there; I'm convinced they'll soon have to release you and I'll mobilize the French Ministry of Foreign Affairs.'

Because I was arrested at JFK in New York (which is under the jurisdiction of the Consulate of New York), then transferred to Wyatt in Rhode Island (which is under the jurisdiction of the Consulate of Boston), representatives of the French Ministry of Foreign Affairs had simply lost track of me, before my sister Juliette alerted them.

It was after this report that the Boston Consulate sent along Mrs L. to visit me. If my sister had heard Mrs L. say that my country would not lift a finger to help me, she would have been aghast.

By phone, I was also able to talk to the children for the first time without telling them where I was. Pierre seemed very surprised when I didn't scold him on his poor maths results for his last exam.

How long are Clara and I going to be able to keep the truth from them? I must get out of here, for my wife, my children and my own sanity. If I stay too long in Wyatt, I'll lose it. I can no longer bear the infernal noise of inmates fighting and their tales of money, cars, drugs, hookers.

# 15

# The US penal system seen from Wyatt

That's it, I don't want to see or hear them any more, and yet I must learn to live alongside them. Contrary to the advice Chris the not-so-paranoid gangster gave me, I decide to tell two or three inmates why I am here. It's hard for me to remain inconspicuous. I don't really look like a narcotics dealer or a gangster. If I wasn't, they would have assumed I was a paedophile: the absolute worst horror! So without revealing everything – I am still very wary of snitches – I tell the bare outlines of my story.

My cell mates all agree that the prosecutors will exert pressure on me to plead guilty. Even under the inequitable circumstances of not knowing what the judges will retain or remove from my prosecution file, I will have to agree to make a deal with the DOJ. Otherwise, I can kiss my freedom goodbye and expect to rot in here for months or even years to come.

In Wyatt, justice does not resemble the justice you see in the movies. While we are all bombarded with American series or films, in awe of a system whereby, during a spectacular hearing, an experienced defence lawyer pleads heroically for an accused, giving the impression that the cause of the weakest has been heard and defended, the reality is quite different, since in criminal matters, there is almost no organized trial. In 90 per cent of cases, the accused prefer to forgo a trial. The reason is simple: the defendants must bear the full exorbitant cost of a defence, and only the wealthiest can afford the services of a lawyer.

Indeed, in the American system, the accused are faced with a prosecutor who investigates exclusively for the prosecution. Unlike the French investigating judge, who investigates both for the prosecution and the defence.

As a result, the accused are forced to pay out of their own pockets for document analyses, counter-expertise, or the search for testimonies that could weigh in their favour. In financial matters, this often means studying tens or hundreds of thousands of documents. Very few indicted individuals can pay the cost of defence counsel – which can run into tens or hundreds of thousands of dollars – or use the services of a private detective to conduct a counter-investigation. Conversely, the US attorneys have the instruments and a significant number of qualified lawyers at their disposal, since, in contrast to France, the American justice system is well heeled.

Thus, from the outset, there is a mismatch in the resources available for carrying out investigations. If, in addition, the accused is in pre-trial detention, and therefore has very limited access to his or her counsel, it is virtually impossible for him or her to be heard, all the more so if he or she is detained in a maximum-security detention facility such as Wyatt.

Admittedly, in criminal matters, at federal level US attorneys can only prosecute cases after obtaining the agreement of a Grand Jury (a group of between sixteen to twenty-three citizens selected at random). This mechanism is supposed to prevent abusive indictments. However, in reality, things are quite different. According to departmental statistics from the American judiciary in 2010, the Grand Juries only opposed 11 of the 162,351 cases presented to them. Finally, if an accused should decide to go to trial anyway, he or she is faced with a judge who has much less latitude in his discretionary power than in France. In the United States, there is a system of minimum sentences, and above all a very restrictive scale of sanctions known as the 'sentencing guidelines', which I will soon discover and which regulate the judge's role, considerably reducing his leeway.

The detainees are therefore totally at the mercy of prosecuting attorneys, who wield huge powers and hold all the cards for making them plead guilty. This results in a DOJ success rate in criminal prosecution matters worthy of a former USSR election result, i.e. 98.5 per cent! Meaning that 98.5 per cent of accused are found guilty.

To achieve the prosecutors' objectives, the US judiciary is prepared to let its quarry stew for as long as it takes. At Wyatt, some inmates have been waiting for a deal for two or even five years. They refused the initial offers made by the prosecutors because they found the sentences too harsh. They also rejected the second offer. And they are now awaiting the third, with no guarantee for their future. It's psychological torture, which has resulted in many inmates suffering from mental and physical disorders.

One of the individuals in Wyatt whom I was closest to, nicknamed The Transporter, whose job was to collect money from the mafia in New York and to transport cash to Las Vegas by private jet where it was laundered, was initially offered a deal of twenty-seven years in jail. Having refused it, and after serving twelve months in jail, the prosecutor offered him a new deal of a fourteen-year sentence. He refused again. Then, after another year of pre-trial detention, he finally signed a guilty plea in which the prosecution undertook not to propose a sentence of more than seven years to the judge. Finally, The Transporter was sentenced to five years, which is exceptional because, in the majority of cases, judges follow the prosecutors' recommendations.

Others are not so lucky, and suicides are frequent. It all depends on how high a tolerance threshold an inmate has.

To avoid losing a case, prosecuting attorneys are party to all kinds of schemes. They even encourage the accused to cooperate by snitching on their accomplices, even in the absence of any material evidence. This penal system is totally irrational and completely twisted. It prompts extreme behaviour. Everyone's priority is to save their own skin, as did the credit card fraudster who sold his wife down the river. She consequently got an eight-year sentence while he got away with only two years. When such cases become public, these inmates are often placed in isolation for fear of reprisals from other prisoners, who abhor snitches.

As for the US lawyers, they are used to and happy to work with this system. Most of them start their careers in the prosecution services as assistant prosecutors or legal assistants, before joining a large law practice. The vast majority of them will never plead in a criminal trial. They are not really defence counsel as we understand the notion in France, but first and foremost negotiators. Their key mission is to convince their client to agree to plead guilty. And then they try to reach the best possible 'deal' with both sides. In their discussions with prosecutors, defence lawyers use a scoreboard system, the famous 'sentencing guidelines'. An outrageous system that I had to learn to work with.

# 16

## My sentencing guidelines

I need to get a thicker skin, a rhinoceros hide to protect myself from this almighty diabolical machine. But how can I remain positive after Liz Latif (who after my arrest had retrieved my professional belongings: phone, computer, iPad ...) informed me that I am banned from accessing Alstom's computer files, that I will no longer receive any mail from the company, that my iPad has been disconnected and my mobile phone subscription has been cancelled. They have cut all ties with me. Guilty or not, I'm no longer a viable option for them and must be ditched without delay.

This was to be expected, but it kills me to think that suddenly I no longer exist for them. Twenty-one years of good and loyal service wiped out. I have to get a grip and fast, because soon I have a momentous decision to make: whether or not to start negotiating with the prosecution.

The prosecutors offered me an appointment for 5 May, three weeks after my incarceration began. They must think I have stewed long enough and that I'm ready. Stan explained to me what is at stake in this proceeding. I noted each of his words in pencil on my A4 notepad. Since the beginning of my imprisonment I have written down every detail of the day, from the menus in the canteen to the insults hurled by the prison wardens, not to mention the stories told to me by the other inmates within the confines of my cell. I naturally record all the stages of the procedure on it. So here is what Stan told me that day.

'The prosecutors are offering you a "reverse proffer". In other words, a confidential interview that serves as a pre-negotiation. They will show you some of the evidence they have accumulated against you to force you to plead guilty. This will avoid them having to hold a trial and will of course enable them to put additional pressure on Alstom's management, to make them plead guilty and fully cooperate with the investigation. Alstom will then be faced with a gigantic fine.'

'So what's in it for me?'

'In exchange for your guilty plea, they will drop some charges, which will reduce your prison sentence. You are currently being charged on ten counts. If we manage to negotiate well, we can hope that they will retain only one charge, i.e. conspiracy to commit corruption, and you will then incur a maximum sentence of five years like Rothschild. Assuming, of course, that everything runs smoothly and that the judge accepts your guilty plea . . .'

'And if I refuse?'

'Well, I would advise against it. They hold the trump card, i.e. their two witnesses: Sharafi the first consultant who's already told them everything, and who's accusing you of knowing about the bribes he paid to the Indonesian parliamentarian E. Moeis. And then there is David Rothschild, who says pretty much the same thing. Moreover, Sharafi and Rothschild have already negotiated with the prosecutors.'

'Precisely, just what are their words worth in such circumstances?'

'I think it may be enough to convince a jury. If you go to trial, you're playing Russian roulette with your future.'

'Maybe, but other than these "witnesses", they have no real evidence against me. The emails they refer to in the indictment do not directly implicate me. If we go to trial, I think I have a real chance of being acquitted.'

'It is precisely these emails you need to worry about! Yesterday, the prosecutors sent us a complete copy of your file: eleven CDs containing at least 1.5 million exhibits. Most of these are email exchanges between numerous Alstom executives over a period of fourteen years. There are also a number of recordings made by the "mole", i.e. the FBI informant. The attorney told us that you can't be heard on the tapes, but we don't know precisely what they contain.'

'So we need to analyse them Stan! This is the priority before making a decision.'

Stan takes offence.

'I don't know if you can imagine how much work this entails. It's a mammoth task. One and a half million exhibits! It will take at least three years to get through and will cost several million dollars in fees.'

This is exactly how their perfectly oiled trap is designed to work.

The machine is formidable, and those who pull the strings are never beaten. In a nutshell, my only hope of being released is to plead guilty. Otherwise, I risk being detained for a very long time, awaiting trial. My cell mates were right. Whatever your initial intention, the prosecutors always succeed in getting you to negotiate ...

With 5 May looming, my brain is nothing more than a merry-go-round of suppositions, calculations, and pros and cons ratios.

When the fateful day arrives, the confines of prison bring me back to reality. Here I go again, shackled like a convict, shoved around with eleven other inmates, crammed into an armoured vehicle en route to the New Haven Court, where prosecutors are about to give me their 'reverse proffer'.

I enter the courtroom and join Stan and Liz, and David Novick, the Connecticut district attorney prosecutor who, on two occasions, vigorously opposed my requests for bail. At his side is another prosecutor whom I have never seen before, and whom I would rather never have met: Daniel Kahn, who has made a special trip from Washington. He is a federal prosecutor inside the DOJ Criminal Division's Financial Fraud and Public Corruption Unit. Young, ambitious, brilliant, this Harvard graduate, a specialist in Foreign Corrupt Practices Act cases, has earned a sound reputation in his criminal prosecutions against white-collar offenders. He even won the award for best assistant attorney.

In the New Haven courtroom, he immediately fires off all his charges. Essentially four receipts – which I have never seen before – from bank transactions made between 2005 and 2009 from one of the accounts of our lobbyist Sharafi (our first 'consultant' in the Tarahan case) to an account held by a relative of Emir Moeis (the Indonesian MP) for a total of about $280,000. He claims that this is evidence of bribery. At least that's what Sharafi will say if I go to trial. Considering the long-established business relationship between Sharafi and Moeis (according to Sharafi they were even co-investors in joint ventures in Indonesia), these payments do not come as a surprise to me. In what way are they meant to be linked to the Tarahan power plant contract? And even if they were, I was never informed by Sharafi or anyone else of these wire transfers, let alone their amounts. That said, I prefer to remain silent. I turn a deaf ear, as Liz Latif strongly advised me to do, shortly before the hearing.

'Above all,' she'd said, 'whatever happens, Fred, you do not show anything, even if they try to throw you off balance. You don't move a muscle.'

So, I stay out of trouble. However, upon each of Kahn's words I feel like a wretched fly stuck in glue, which despite struggling, sinks inexorably into it. In their statement, Kahn and Novick qualified me as a simple 'link in the chain'.

The only thing that matters to them, they claim, is to trace the perpetrators up the ranks to Alstom's CEO, Patrick Kron. Our session ends after half an hour. The two prosecutors did not question me. They just wanted to show their muscle at this stage.

The ball is in my court and I can't afford to dither too long, because, as Stan Twardy tells me at the end of this hearing, time is my number one enemy.

'There is still something they haven't told you,' explains Stan, 'which makes your situation even more tricky. They have just indicted a third Alstom executive, Bill Pomponi [Rothschild's successor, retired for many years]. So, they're playing for time.'

'I don't understand any of this, Stan. How does Pomponi's arrest impact on my situation?'

'Well, they will most certainly try to strike a deal with him too. And if Pomponi accepts to plead guilty before you do, and gives fresh information to the prosecutor, your bargaining position will be undermined and you will no longer be able to negotiate a guilty plea at the same level. They may even lose interest in you and "forget" you in custody until you face trial. You really need to decide fast. You have two or three days, no more!'

'But on what basis do I decide, Stan? All this is diabolical. Should I plead guilty, Alstom will cease to pay your legal fees and cut me loose. If I plead not guilty, I risk a one hundred and twenty-five-year sentence after a trial. And I only have two days in which to make this decision because Pomponi may "steal" my spot. All this when I cannot even access my file because it is too extensive, containing one and a half million exhibits, and it would cost a fortune for your firm to analyse them. Are you serious?'

'This is as serious as it gets, Frédéric. That said, don't focus on the hundred-and-twenty-five-year sentence; that is purely theoretical.'

'So what am I really risking? Are you finally going to tell me? What are these famous *guidelines* that the other prisoners told me about?'

Liz raises her hand as if she were dealing with a capricious child.

'Calm down, Mr Pierucci. And listen carefully.'

What she goes on to tell me proves to be one of the most outrageous things I have heard.

'We have commissioned a report,' she says. 'An FCPA offence is worth twelve points. Second, the amount of the gross margin earned by Alstom on the contracts in question, i.e. $6 million, must be taken into account, which equates to an additional eighteen points. Finally, the bribes were paid to a parliamentarian, and therefore to an elected official, which is an aggravating circumstance. So we have to add four points.

'Then we have to add two points because there were several payments, which, according to the DOJ, equates to several bribes. This brings the total to thirty-six points. If the prosecution had deemed you to be the ringleader of the conspiracy, four more points would have been added, but that is not the case, so we stay at a total of thirty-six points.'

'What are you driving at, Liz, with your points calculation? What matters to me is how long I'm going to spend in jail!'

'I'm getting to that. Then you have to transfer your thirty-six points to a double-entry table, with horizontal lines, i.e. the number of times you have already been convicted, and therefore vertically, the points corresponding to the offence you are charged with. In your situation, we are faced with a penalty range of a hundred and eighty-eight to two hundred and thirty-five months. If you go to trial and are convicted, if the judge follows this scale – which is almost always the case – your sentence will be at least fifteen years and eight months and may even increase to nineteen years and seven months.'

'But Liz, how can the DOJ use a calculation formula like this? First, why are you taking into account the margin achieved by Alstom to calculate my hypothetical sentence? This was not personal gain but went straight to the company, and therefore into the shareholders' pockets. I didn't even make one dollar; there was no kickback. In effect we are treating an employee who acted on behalf of his company for no personal gain and one who obtained personal gain in exactly the same way.'

'Absolutely. The DOJ considers that the employee acted to keep his job, get a promotion, or earn a bonus. Therefore, in its view he made personal gain.'

'Then, in this case, this supposed gain would have to be quantified.'

'Look, there's no point in arguing. You're not going to rewrite US law! The calculation rule is what it is, period.'

'And the payment? We are dealing with a single consultant's contract. Even if Sharafi's remuneration was paid in several instalments, why two further points for several payments?'

I can see Liz turning red and about to explode. Stan intervenes and in a cursory tone reminds me that the FCPA was created in this way, and that 'there is no point in arguing'. I am speechless.

When I return to Wyatt, I am so distraught that I am incapable of thinking straight. Haggard, I wander into the common room, the only place in the prison where prisoners can meet in a shared living space. I observe inmates playing chess. One of them excels. His moves are unparalleled. Then, after winning his game, he joins me and tells me of how he came to be caught with five hundred kilos of weed hidden in his house. When he was thirteen years old, his mother, a prostitute, fled the marital home, then his father abandoned him in a wood. *Hansel and Gretel* springs to mind. He had to steal to survive, then gradually made it; well, almost. He started growing hemp.

In my head, I start converting the five hundred kilos of cannabis into points, then the points into years of prison. Since he must be a re-offender, there is also the risk of repeat offences. Suddenly I stop, as dizziness grips me. 'To survive in prison, Mason my cell mate keeps telling me, close your eyes and breathe deeply. Keep breathing and then just live.'

For the first time, this evening of 5 May 2013 in Wyatt, I join the other prisoners to watch TV in the refectory. Exceptionally, a recorded version of a European football match is being broadcast: the Champions League final. Against all odds, Bayern Munich beat Barcelona 3–0.

# 17

# A pod

The next day, I can hardly believe it when a bald little man of about seventy-five, with most of his teeth missing, greets me with a feisty 'Bonjour Monsieur!' in a perfect French accent. The prison administration has just transferred me to A pod.

'Hi Frenchie,' he continues, 'I'm Jacky, but here I'm known as the old man. We've been expecting you for ages!'

Looking at my amazed face, Jacky explains to me that he has 'some connections' in Wyatt's administration and that he was behind my transfer to A pod.

'Since you arrived at the facility, I've been pleading with them to send you here. I'll be able to practise my French with you.'

Jacky is one of the quirkiest characters I have ever met. A legendary figure of organized crime, he is one of the few survivors of the French Connection, the famous Marseilles drug-trafficking ring that supplied the American mafia with heroin from the 1930s to the 1970s. Jacky first began his career as a gangster in New York City, in the Bronx. He was arrested in 1966, sentenced to five years, and then arrested again in 1974. He managed to escape to France, heading for Marseille. The Narcotics Brigade stopped him in his tracks again in 1978.

Extradited to the United States, he served a twelve-year sentence in a detention facility. However, as soon as he was released in 1997, he started selling heroin again and was sentenced to a fourth and then a fifth term. A total of thirty-six years in prison, four of which were spent in France. That's how he ended up in Wyatt. Needless to say, with such a background he has a lot of fans behind bars. In fact, he knows everyone and, most importantly of all, he is known and respected by all the inmates. He is the true captain of this ship.

Despite his background, he is also very friendly and caring. His fascination with France was the reason he wanted to be placed with francophones such as Alex the Greek (a graduate of the Marseille business school and former BNP employee), and a Canadian of Greek origin, the famous Transporter. All three of them treat me like royalty. They give me coffee, sugar, powdered milk, a radio, a mirror made of aluminium foil, new sneakers (while I await the pair I ordered from the commissary) and, best of all, a decent pillow and second mattress.

Unlike D pod where I had been living since my arrival, in A pod there are no cells. It is a vast dormitory housing fifty-six people. The inmates are grouped into small cubicles holding four people, separated from each other by five-foot-high partitions. Each cubicle contains two bunk beds and is no larger than ten square yards. Aside from the overcrowded conditions, the toughest part to cope with is the lack of natural daylight. It only reaches the pod via three tiny vaults, covered with an opaque film. So we permanently live under the glare of fluorescent lights, with one in two lights left on at night to identify troublemakers.

Those who, like me, have the upper bunk, must adjust to sleeping with these lights placed just twenty inches above their heads. It took me three nights to finally be able to sleep, my eyes desperately focusing on the ghastly yellow flaky paint walls. A pod has not been renovated for nearly ten years. The bottom line is always to reduce prison operating costs to increase profitability.

In this ward, not only are the showers communal but so are the toilets. Five WCs are aligned against a wall, separated on each side by a low three-foot wall but with nothing in front.

Within a few weeks of their incarceration, the prisoners are categorized and placed in their final quarters, i.e. the pod appropriate to their age or their level of dangerousness. Pods for 'gangs', pods for 'workers', etc. A pod is designated for prisoners over forty years of age, who are normally less disruptive. It houses a large Latin community (Dominicans, Jamaicans, Mexicans), some Asians and, strangely enough, nine Greeks or Americans of Greek origin. I am the only Frenchman.

Most of my cell mates were arrested for murder, robbery, drug trafficking, or just credit card fraud (a Vietnamese speciality here). In this run-of-the-mill spectrum of crimes and misdemeanours, there was only one FCPA case, an offence that no prisoner had ever heard of, before I arrived in A pod.

This quarter is supposedly reserved for the most placid individuals. I dread to think what the other pods are like. The reason I say this is that in A pod, arguments, thefts in the dormitory, and drug and medication trafficking are commonplace. Not a week goes by without an inmate being placed in solitary confinement, which can last from a few weeks to several months. The first rule to avoid trouble is never to stare at or touch another inmate. No shoulder slapping, no hand shaking, and when waiting in line for meals, you must not touch or even brush past anyone. The slightest gesture is automatically interpreted as an act of aggression.

The walk yard, adjacent to the pod, even though tiny (half the size of a basketball court) is a haven of peace. In fact, the prisoners have their own code of unwritten rules. From 8 a.m. to 11 a.m., the courtyard is for people doing gymnastics or walking. The rest of the morning and afternoon it is transformed into an area where prisoners play Basque Pelota using their bare hands. Then finally, between 8 p.m. and 9 p.m., walkers are allowed back into the yard.

While I'm settling into my new cubicle, my new cell mates also introduce me to another ritual that governs the sequencing of events in the common room. At mealtimes, inmates must always sit at the same table and in the same place. Between meals, for whatever reason, the places change, but the ritual remains: each hour of the day corresponds to a positioning that the newcomer must learn. It's quite confusing at the beginning, but then you learn to comply with this unwritten rule.

I received the first pictures of the children in my new 'room'. Clara chose the nicest. They are laughing to their hearts' content and that does me a world of good.

Alex and The Transporter have not seen their wives and children since they were jailed; i.e. fifteen and twenty-two months respectively. They told me that before their arrest their relationships were solid and then, gradually, they got more complicated.

Yesterday, by phone, my sister Juliette and Clara told me that my father was planning to visit me in the United States. I did not want him to make this exhausting trip. The prospect of seeing him for an hour behind bullet-proof glass and only being able to talk to him through a telephone handset saddened me. I do not want to put him through such humiliation at his age. I also fear being crippled by shame too. Though if I were in his shoes, I would of course have jumped on a plane to be at my son's side in jail, calming my own anxieties in the process. I know I'm depriving him of reassurance and peace of mind, for he must be climbing the walls wondering what he can do to help me.

Tomorrow, a very important decision awaits me: whether to plead guilty or not guilty. I called Clara to explain the dilemma as it will have a huge impact on her and the children. The only thing that matters to her is that I am a free man, no matter what the consequences are for the family and for my work; there is no price on freedom.

I still have all night to think about it. Bearing in mind at least one piece of good news: Linda (an American friend we met when I was based in Windsor, Connecticut) has agreed to post bail on her house so I can be released. This is tremendous news, but will it suffice?

# 18

## Cut loose by Alstom

My first night in the A pod was tough. Fifty-six of us in a dormitory. There are those who snore, those who pass wind, and those who discreetly or not so discreetly masturbate, endless visits to the toilets, plus the guards doing their noisy rounds.

After breakfast, I call Stan. In fact, I haven't reached a decision. It all depends on what he says to me about the actual sentence.

'Fred, I have some good and some not so good news for you,' Stan begins. This sets the tone.

'Start with the bad.'

'Since your arrest, Alstom seems to have adopted another tactic. Up to now, your company has shown itself to be very reluctant to cooperate. However, they have now started to provide the DOJ with the documents requested. Prosecutors have received tens of thousands of additional documents, including three thousand that mention you. I get the impression that . . .'

'You get what impression, Stan?'

'I'm not sure, but I wonder if some people at Alstom are not trying to exploit this situation to frame you for contracts other than Tarahan, to cover themselves.'

'But how would they benefit? I don't see why they would do that. It could backfire on them.'

'Now listen, Frédéric. Your arrest has made them realize that this is no laughing matter. They have now grasped that they must proceed to the checkout and settle a mega fine. But above all, they – that is the CEO and other senior officials – live in fear of individual actions being taken against them. So damage limitation is in their interests, meaning best to sentence the one already behind bars.'

'If they are intent on playing that game, me too, I can also name dozens of executives who were involved. In fact, the entire management of Alstom itself!'

'I know, Frédéric, but if Alstom is willing to cooperate fully and pay the fine, the DOJ will be more sympathetic to your managers. The main point I want to make is that this may delay the process, as the prosecutors will want to cross-examine you on the documents sent by Alstom concerning you.'

'Is that going to take a long time?'

'Not that long. Of the three thousand documents, there are perhaps a few hundred that are relevant. If you agree, they will send Wyatt some CDs that you can review on a prison computer and then they will schedule some meetings at the New Haven Court where they will cross-examine you on these documents. If they are satisfied with your answers, only then will they agree to drop the charges against you.'

'And, alternatively?'

'There's no alternative. Or you remain in prison at Wyatt awaiting trial.'

'Please reassure me. Can we still ask for my release on bail? An American friend of ours, Linda, has agreed to put up her house for bail.'

'Yes, we can. But I'm not sure after this shift that Alstom will still be willing to pay for the guards and the rental of the apartment. If you have to pay for it yourself, the cost of guarding the apartment 24/7 will be exorbitant. Besides, the prosecutors will not agree to your release until all the documents have been examined.'

'I thought you had some good news for me too?'

'I do: the good news is that if you agree to review these

documents and answer their questions, they are prepared to recommend a six-month prison sentence.'

At that moment, I was relieved to hear that figure. And that was exactly the desired outcome. They frighten you with lengthy jail sentences if you go to trial, to force you to ultimately plead guilty in exchange for a very reduced sentence. Hey presto, I'm now part of their inferno from which virtually no one escapes.

We have been talking on the phone for more than fifteen minutes now and the line may be disconnected at any time, each conversation being limited to twenty minutes. I still have dozens of questions about the terms of the guilty plea, the prosecutors' interrogation sessions, the guarantees for the six months, and the charges that will be dropped, but Stan is urging me to hurry up and make a decision.

'You have to make a decision. The prosecutors have no doubt offered the same deal to Pomponi, and if he agrees, it will no longer be six months for you, but much longer, and perhaps even no offer, except to plead guilty on all the charges.'

I attempt one last question.

'What guarantees do I have that the judge will comply with the prosecutor's request for a six-month sentence?'

'None. The judges decide themselves. However, in the vast majority of cases, they follow the prosecutors' recommendations, especially here in Connecticut.'

'Stan, if you can confirm to me that the prosecutors will suggest a six-month sentence to the judges, then you have my green light.'

I ought to be relieved. I have finally made the decision. In spite of that, I am besieged by thousands of other questions. How will Alstom's Board react when it learns that I have pleaded guilty? If the company ceases to pay my salary and the cost of living in Singapore, what will become of my wife and children? They will not be able to stay in Singapore and will surely have to return to

France. Admittedly I will be free, but I will be forced to wait for my sentencing alone, without being able to work in the United States. Would it not be more appropriate to divorce? I could leave the house in France to Clara. Then at least she could start over again.

# 19

# Back to court at New Haven

It went awry, horribly awry. From the furious look on the faces of the prosecutors Kahn and Novick, the answers I gave were not the ones they had hoped for. They call time out to 'give me time to think'. My lawyer uses this moment to yell at me and slam my performance. He is furious.

'What the hell were you thinking? Why did you deny it?'

'I did not deny anything. All I did was tell the truth, believe it or not. Yes, Alstom paid a 1 per cent commission to our consultant Sharafi. But no, I never discussed with him the possible payment of a bribe to the Indonesian parliamentarian Emir Moeis, who, moreover, had no official role in the award process of the project.'

'But you must have known that it involved a kickback, didn't you? You must have known it was a possibility! Besides, Sharafi has already testified. And he blew the whistle on everyone at Alstom. So stop fooling around and tell them what they want to hear; otherwise, I warn you, the process will be halted and you will be sent back to jail right away and you can kiss the six months goodbye.'

'So I have to lie, Stan. This is just completely insane. If they want a confession, I am quite willing to admit that yes, when Alstom recruited Azmin, the second consultant, with the approval of top management, I knew there was a real risk of potential bribes being paid by him. There was therefore little doubt about why Azmin was hired. However, when Sharafi was first recruited, there was no question of asking him to pay bribes. At least that's what he told me.'

'Okay Frédéric, but right now they don't give two hoots about the second consultant. They have built the whole case against you from the testimony of the first consultant Sharafi. And they're not going to change their strategy now.'

'So what do I do?'

'Well ... Listen very carefully to what I'm about to tell you ...'

My lawyer pulls his secret weapon out of his sleeve to convince me to change my story while avoiding lying, and not to ruin his negotiation, i.e. 'wilful blindness'. In other words, I bury my head in the sand. Even if the emails do not prove my personal involvement, even if I did not know the destination of some of the payments to Sharafi, I am still guilty of intentionally keeping myself unaware of facts that would render me liable or implicated. I 'wilfully turned a blind eye'. And for the DOJ, the consequences are the same.

I memorized the statement that Stan was whispering to me, then I went before the prosecutors and recited everything to them. What else could I do? Novick and Kahn were delighted with my U-turn.

I later find out why they were not interested in Azmin, the second consultant, but in Sharafi, the first, although corruption was much more likely on one side than on the other. When Azmin was paid in full twelve months after the contract entered into force, i.e. 2006 at the latest, the facts were time-barred (five years in terms of FCPA), whereas with Sharafi, the last payment by Alstom was made in 2009, which was still in the non-prescription period for my indictment in November 2012.

When I got back to Wyatt, after such a tiring day, I called Clara. She split the tasks between herself and my sister Juliette. Juliette, with the help of her husband who is fluent in English, coordinated the legal actions with my lawyers, while my wife was in charge of maintaining contact with Alstom. Or rather she tried to, because overnight I had become nothing less than a black sheep for Alstom. Clara heard the rumours that were circulating within the company. I still can't get over what she found out.

Alstom's legal department, who did not deign to contact me, drew their own conclusions. They believed that regardless of my initial intentions, I would be forced to plead guilty, and lo and behold they were right. They knew that Rothschild and Sharafi had 'cooperated', which they cited as evidence. There was even talk that they had both managed to obtain 'whistle-blower' status and that, as provided for by the US Dodd–Frank Act,* they would be rewarded for their 'betrayal'. They could be paid between 10 and 30 per cent of the fine imposed on the company they exposed (a former UBS employee hit a $104 million jackpot).

The rumours surrounding Rothschild are false. To the best of my knowledge, he will never be paid by the Department of Justice, which induced him to cooperate by promising him a reduced sentence.

A state of panic engulfed the company after my arrest. Everyone in the office desperately tried to recall their numerous conversations, all wondered whether a 'mole' had been present that particular day, whether he wore a microphone, and above all whether they had said anything incriminating. They all lived in fear of the FBI turning up at their offices. Indeed, some former employees even asked for a lawyer to be assigned to them from the company's legal department should they encounter difficulties. Many now considered Patrick Kron, CEO, to be in the hot seat and on the brink of indictment. At the headquarters in Paris it was action stations.

I later learned that immediately after my arrest, the legal department sent an email to about fifty executives whom the DOJ had

---

* Dodd–Frank Wall Street Reform and Consumer Protection Act of 2010.

questioned Alstom about since the beginning of the investigation. Such a warning was highly unusual for Alstom.

'As you know, a judicial inquiry is underway in the United States into alleged corruption in some international projects. An internal audit of the company shows that you have been involved in certain US projects … It is possible that the US authorities may try to cross-examine you if you travel to the United States,' states the document. 'Check with Keith Carr before travelling to the United States for any Alstom-related work.' Management advises likewise, offering a list of recommendations to its executives: 'If arrested, you should know that you have the right to speak or not speak to the investigators. It is your choice and the American government cannot force you to speak.'

Why did they wait till I was arrested before warning these employees, knowing for a long time that Alstom was in the firing line of the US judiciary? And why did Keith Carr reassure me shortly before my arrest that I was safe? I learned later that this memo was prepared before I was arrested but not issued and that I was not in the original recipient list. Why Alstom management did not issue it at that time remains a big question mark. Then why, given these circumstances, was I arrested? Why me? This question has been exasperating me since my arrest.

Did the new role I was about to assume (future head of a joint venture between Alstom and Shanghai Electric) have anything to do with it? This partnership, had it proceeded, would have made the group the global leader in coal-fired power plants.

According to analysts, it would also have heralded a closer global partnership between Alstom Power and Shanghai Electric, posing a serious threat to our major rival General Electric; something the Americans obviously feared.

Until I get the answers to my questions, I can easily imagine the atmosphere of suspicion and anxiety that must prevail at the Paris headquarters and all the group's units abroad. In fact, I wouldn't be surprised if we were shortly to witness a round of musical chairs in the executive committee, aimed at protecting the managers most at risk. Unfortunately for me, my fate is already decided.

# 20

## The evidence

I must stay strong. Emotionally and physically I need to keep a close eye on myself and stay in shape. Here the one commodity we have is time. For the past three days, I have joined a group of three inmates, who work out every morning in the prison yard. I am so physically unfit that I can't even do more than three press-ups in a row. But I hang on in there. I have compiled a complete programme for myself. In the morning I work out. In the afternoon I study the evidence. There is a small room equipped with six computers dating back to Noah's Ark. This computer equipment is made available to prisoners to enable them to consult their criminal files and the evidence gathered by prosecutors.

At the entrance, a member of prison staff gives us envelopes with our names on, sent by our lawyers. Liz has sent me four CDs. I have the right to view them on a screen and take notes, but I am not allowed to print any documents from them. The CDs contain the famous 3,000 exhibits sent by Alstom, but I suspect that the prosecutors have added others that they have obtained by other means. Many are stamped as originating from the Swiss police (in 2011 Alstom had already been investigated and convicted of corruption in Switzerland).

They consist of emails that I sent, received or simply copied on between 2002 and 2011. For the ones dating back the furthest, it is almost impossible to recall the details of specific projects. I rapidly calculate in my head that if I need to read the 1.5 million pieces of evidence that prosecutors have gathered, spending one minute per document, and knowing that I only have access to this room for one hour a day, it will take me sixty-eight years to read the entire file. I find this both preposterous and shameful. Such a justice system violates the most fundamental human rights. The Department of Justice knows that time is on their side, so they purposefully drown the accused in tons of paper. They pursue the same old logic relentlessly, i.e. deprive the accused persons (except of course the very wealthy) of a defence mechanism to force them to plead guilty.

I scrutinize the content of the CDs every day. At times a small voice inside me says that this frenzied reading is probably all a big waste of time. Who knows? If I'm lucky, I may uncover a gem. The one 'lucky strike' that silences my accusers. More than anything, the CDs give me a purpose. The chance to use my brain.

In just a short space of time, prison life has numbed my grey matter. No more watches, computers, iPads, planes, meetings, desks, work projects, i.e. zero mental stimulation. The most riveting part of my new existence is now whether we will be allowed a chicken leg (three times a month) or whether we will be given ice cream next Sunday.

In the A pod, the atmosphere has suddenly deteriorated. There are more and more altercations, fights and violence in general. Two days ago, a big black dude stole my mirror and a Turkish guy who saw him do it immediately stepped in. I tried to calm them down, they started shouting and insulting each other. The whole pod witnessed the scene. The two guys went off to fight in the showers, the only place without cameras, and some armed guards arrived just in time to prevent them from killing each other. The two combatants were put in the 'hole' (solitary confinement), which never seems to be void of an occupant these days. I live in fear of retaliation. The fight started on account of my mirror, right? In this place, violence can strike you at any moment.

Soon after, I learned that the guards had placed a newcomer, a brutish lout, in the cubicle of a little old fellow that everyone knew well.

This new inmate was also a serial rapist. The wardens must have thought that he was unlikely to attack an older inmate, or perhaps they didn't think about it at all. They simply wanted to fill the empty bunks. At night, we heard screams; we guessed what had happened. In the morning, it was too late; the little old man was taken to the infirmary.

For a week now, I have been attending church. The Latinos and of course the Greeks in the pod all go. I hadn't set foot in a church in four years, and that was for my nephew's first communion. The priest preached about forgiveness. To know how to forgive others, and yourself. I don't believe in God, but Christ's message is universal. Maybe there is an underlying reason for my presence inside these four prison walls. Maybe when I get out, I'll have a deeper, more balanced, more authentic existence? Maybe I'll be a better father? A better son? A better brother? A better husband? I've put poor Clara through so much.

At least I can see a little more clearly now, after a month of uncertainty. If I had been released on bail, they would have forced me to remain in the United States where Clara and the children would have joined me.

She had already organized everything. The move from Singapore, finding herself a job, enrolling the children in new schools. She even found an apartment in Boston. She wasted a crazy amount of energy for nothing. My ongoing detention thwarted all these plans. We also had to decide about the start of the school year in September, because the International High School of Singapore, which all four children attend, is in high demand and you have to reserve your place and pay a large deposit at the beginning of May.

Consequently, the family will stay in Asia at least for this school year. This is probably easier for everyone.

This week, I also had a visit from my friend Tom, a Franco-American whom I met when I arrived in the United States in 1999. I met him in the visitation room, where we have to use handsets and are separated by bullet-proof glass. As many visitors come with their families, often with young children, the noise is indescribable, and it is very difficult to hear each other. Anyway, I was very happy to see a familiar face. He is in contact with Clara and has promised to call her back when he leaves to reassure her of my health and mental state. We can talk for one hour, not a minute longer, after which the handset is automatically switched off. All conversations are recorded, which limits the extent of our discussions. I tell him to reassure everybody that I am fine. The time goes by swiftly and I quickly find myself back in the A pod.

Tom is one of the very few friends I have who visits me in jail. Others, whom I was close to when we lived in the United States from 1999 to 2006, do not have the courage to walk through Wyatt's doors, for fear of having their names put on a list by the US authorities, which I fully understand. A few days later, I get a visit from Linda, to whom I am eternally grateful for putting up her house as bail bond so I can be released. This is an extraordinary act of kindness.

# 21

# The worldwide reach of the prosecutors

The prosecutors are not letting me out of their sight. Between mid-May and early June, Kahn and Novick summon me three times to New Haven for questioning. They have embarked on a veritable world tour. They endlessly brandish emails exchanged within the company between 2002 and 2011 about contracts signed or even simply sought by Alstom in India, China, Saudi Arabia or Poland. Their questions are extraordinarily precise: 'What do these initials signify? Why does this person call his correspondent "friend"? Have you met these people? If so, when? And who was with you? Did you use consultants for this project? If so, who were they? What were they paid? And what were the payment terms?'

In the pile of documents they show me, the Indian projects of Sipat and Bahr I seem to captivate them the most. I remember that, at the time, these projects, negotiated between 2002 and 2005, caused considerable bad feeling between the various entities of Alstom, with, on the one side, the two sectors Power Environment (in charge of the boiler) and Power Turbo-Systems (in charge of the turbines) and on the other side, the sales organizations 'International network' and 'Global Power Sales'. The main disagreement was over the choice of 'consultants' between ABB's long-established networks and Alstom's. It became a real battleground. Personally, I never met or was in contact with the consultants they finally selected. Alstom ended up not bidding on Sipat and lost Bahr I on price. End of story. Though not for the DOJ, which in 2013 started delving into the projects lost in 2004–05. Why, I wonder?

The prosecutors fire question after question at me. I try to answer as best I can by sticking to the essentials. But all this was so long ago. All I can focus on is finishing this endless interrogation so that I can be released. The fourth and final hearing with the prosecutors is scheduled for the end of the first week of June. Ordinarily this is just a formality. Dan Kahn and Dave Novick will ask me to repeat my admission of guilt. And at last I will be able to file my application for release, and it will no longer be contested, particularly since Clara has finally raised the $400,000 needed for my bail, and our friend Linda has risked putting her house on the line for me. According to Liz, this should be sufficient. If the schedule has not altered, I am confident that I will be able to leave around 15 June.

My cell mate Jacky, on the other hand, is sorry that his Frenchie will soon be going. He makes me promise, as soon as I get out, to send him a CD of Nicole Croisille, his favourite French singer whom he saw live at the Olympia in 1976 while he was on the run in France. He can still remember her singing 'Tell me about him'. That night I sleep like a log under the fluorescent light dreaming of Paris. Only twenty-four hours to go before my last hearing with the attorneys.

What I believe to be the last session begins. All the documents provided by Alstom have been reviewed. With Stan too, all is set. We have rehearsed in detail the theatre piece that should satisfy the intractable duo, Dan and Dave. Not my favourite people. Then suddenly, at the end, Dan asks to have a private meeting with my lawyer.

They all retire to an adjoining room. So why don't they want me to participate? Did I respond in a way that they didn't like, as in our first meeting? Has Pomponi provided fresh information? Has Alstom produced some new evidence? Are they going to bring new charges against me? Their confabulation goes on and on ...

Suddenly the door opens. Stan Twardy comes out alone and sits opposite me.

'Okay, I'll summarize the situation for you. If you maintain your request for bail, they will oppose it.'

'What are they concocting now, Stan?'

'It's the same old problem: they want to make an example of you. Sharafi, who was arrested first, obtained total immunity. Rothschild, who was the second to be arrested, was able to negotiate with them. Unfortunately for you, you are in third place in the race and higher up in Alstom's hierarchy. So, using their logic, you should pay a higher price. Whatever the kind of guilty plea you manage to negotiate with them, they want you to stay in jail for six months to prevent you from contacting anyone outside, notably Alstom.'

'This is completely absurd. If they arrested Rothschild before me it's simply because he is a US citizen, and lives in the United States, while I happened to be abroad.'

'For once, I agree with you Frédéric, but they've got us cornered. Either you accept it and do your six months, or we maintain tomorrow's bail hearing, but your chances of winning are now very slim.'

'Look, I'd like to discuss it with Clara.'

'Sorry Frédéric, that's not possible. You must decide now. Either we maintain the hearing, or we push it back again. They have given you ten minutes to decide.'

Ten minutes. I once again apply the method recommended by Mason, my former cell mate from D pod. I take a deep breath. After all, it is straightforward. If they want to put me away for six months, no matter what I do or say, I won't be able to avoid it. As I have already spent nearly two months in prison, I would thus only have another four to serve. Either I serve them now or I will have to serve them after my sentencing. This time, they most probably won't place me in a maximum-security jail like Wyatt, but I might as well get it over with now. I can understand Stan's reasoning. I will return to Wyatt for four months as sought by Dave and Dan. I am still baffled by their unrelenting pursuit of me. It is clearly not just my guilty plea. There must be something else at the source. It's going to take me some time to find out what.

# 22

## The FCPA

I am obsessed with the FCPA. This acronym is the reason for my incarceration in Wyatt. I have little knowledge of this legal mechanism. Stan and Liz have given me some information on it, but despite my repeated requests, they have only provided me with very brief documentation. Fortunately, Clara was able to find an 800-page study by an American firm that listed all past corruption prosecutions. Since receiving it, I have been frantically scrutinizing all previous cases and comparing them with my own. Over the next few months, I become an FCPA junkie and an expert on it. However, in the spring of 2013, I haven't yet got to that stage, I have just started.

I discover that this law was enacted in 1977 following the famous Watergate scandal. By investigating the political scandal that led Richard Nixon to resign (a break-in at the Democratic Party headquarters), the American judiciary uncovered a gigantic system of dissimulated financing and bribery of foreign public officials. Four hundred American companies were involved. The US Senate committee in charge of the investigation revealed in its conclusions that members of the board of directors of Lockheed, one of the largest US defence corporations, had shelled out tens of millions of dollars in bribes to politicians and public company executives in Italy, West Germany, Holland, Japan and Saudi Arabia to sell its fighter planes. Lockheed admitted that it paid Prince Bernhard, husband of Queen Juliana of Holland, more than a million dollars to help sell its F-104s, which at the time were competing with the

French Mirage 5 aircraft. In response to this great scandal, Jimmy Carter's presidency enacted a law prohibiting American companies from bribing 'foreign public officials' (civil servants, political leaders, persons delegated with a public service mission). Two agencies are responsible for enforcing this law: the DOJ, in criminal cases, which prosecutes companies and individuals who violate the law, and the SEC (Securities and Exchange Commission), in civil cases, which prosecutes companies suspected of falsifying their accounts (and thus misleading investors) to conceal expenses related to bribe payments. In general, the SEC only intervenes if the company is listed on a US market (NY Stock Exchange, Nasdaq).

However, as soon as it came into force in the late 1970s, the FPCA, the world's first ban on bribing foreign public officials, was vigorously contested by major American players. They considered that this new legislation could severely handicap them in export markets (energy, defence, telecommunications, pharmaceuticals, etc.). Indeed, the other major economic powers, particularly in Europe (France, Germany, Great Britain, Italy, etc.) had not yet adopted similar measures. On the contrary, they continued to get away with it, using the services of 'consultants' in countries where corruption was rampant. Some nations, such as France, had even instituted an official system for companies to declare bribes to the Ministry of Finance in order to be able to deduct them from corporate income tax. This practice persisted in France until the year 2000. Different era, different methods.

As a result, the American authorities, not wishing to endanger their own export industry, were less than zealous in implementing the FCPA. Between 1977 and 2001, the DOJ only sanctioned twenty-one companies, often second-tier ones. This doesn't even equate to one per year.

However, America's corporate elite were not happy that this new law was being 'stalled'. By this time, they had figured out the potential benefit they could derive from it. However, for this to happen, their global rivals would have to be subjected to the same treatment, and in 1998, they won their case. Congress then amended the law to give it extraterritorial reach. Henceforth, the FCPA also applied to any foreign company.

The United States believes it has the right to prosecute any company that has concluded a contract in US dollars, or even if emails – considered as international trade instruments – have simply been exchanged, stored (or transited) via servers based in the United States (such as Gmail or Hotmail). With this law, the Americans have succeeded in pulling off a neat sleight of hand. They have transformed a law that could have weakened their own industry into a formidable instrument of underground economic warfare and intervention.

From the mid-2000s onwards, the DOJ and the SEC continued to test the limits of this extraterritoriality, not even shying away from considering foreign doctors as 'public officials' – due to their public service delegation – in order to prosecute international pharmaceutical companies.

While in 2004, the total fines paid by companies under the FCPA were only $10 million, in 2016, they skyrocketed to $2.7 billion. A great leap forward made possible by the promulgation of the Patriot Act in 2003, following the attacks of 11 September 2001, which gave American agencies (NSA, CIA, FBI) the power to snoop on foreign companies and their employees on a massive scale, under the guise of the fight against terrorism, which, for the purposes of obtaining public procurement contracts, is totally unfounded in most cases. Naturally, the beneficiaries of corruption are first and foremost unscrupulous officials, or political parties, rather than Daesh or Al-Qaeda. This espionage was first exposed when the PRISM scandal broke in 2013, following Edward Snowden's revelations.

It then became a global reality that the US digital giants (Google, Facebook, YouTube, Microsoft, Yahoo, Skype, AOL and Apple) were also prepared to share information, when required, with US intelligence agencies.

It doesn't end there, however. Not content with possessing exceptional intelligence resources, the American authorities campaigned relentlessly to the OECD for its member countries to adopt their own anti-corruption laws as well, which France effectively did from May 2000. The difference being that European nations neither have the means nor the ambition to enact extraterritorial laws. This has left them trapped. By acceding to the OECD Anti-Bribery Convention, they have de facto allowed the United States to prosecute their companies, without having the legal means to retaliate and prosecute US businesses in return. It is a malignant mechanism that we have all succumbed to, or rather the OECD member states have.

Indeed, China, Russia and India, which are not part of the OECD, have not yet adopted anti-corruption laws targeting their own exporting companies.

I am not suggesting that we should not fight corruption; quite the contrary. The staggering sums that end up in the pockets of rogue officials, despots or influential members of ruling families would be of far greater use to poor or underdeveloped countries. Corruption is a cancer. According to a World Bank estimate, in 2001–02, $1 trillion was used for bribery. This represents 3 per cent of global trade over the same period. It goes without saying that this money could and should have been used for the construction of schools, hospitals, dispensaries and universities in many countries. This worldwide scourge must be tackled. But let's not get into the wrong battle.

Under the banner of moral virtue, the FCPA is first and foremost a formidable weapon of economic domination. The question we should ask is, has corruption decreased significantly between 2000 and 2019? The answer is no. One thing is certain, however. This law is heaven-sent for the US Treasury, a real little goldmine.

The penalties imposed, which remained modest for several years, skyrocketed from 2008 onwards, with foreign companies contributing the most. Between 1977 and 2014, only 30 per cent of the investigations (474 in number) targeted non-US companies, but they paid 67 per cent of the total fines imposed. By mid-2019, of thirty fines exceeding $100 million, twenty-three were inflicted on non-US companies.* Some are German: Siemens ($800 million), Daimler ($185 million), Fresenius Medical ($231 million). Some are French: Total ($398 million), Technip ($338 million), Alcatel ($137 million), Société Générale ($293 million). Some are Italian: Snamprogetti ($365 million); Swiss: Panalpina ($237 million); British: BAE Systems ($400 million), Rolls Royce ($170 million); Russian: MTS ($850 million) ; Japanese: Panasonic ($280 million) and JGC ($219 million). An impressive international scoreboard for an 'American' law.

Notwithstanding the above, some American companies have also been in the firing line. I am surprised, however, that in more than forty years of enforcement of the FCPA, the DOJ has never found anything reprehensible about the practices of US oil tycoons or its defence giants. How is it possible that these US giants are the only enterprises that manage to secure contracts in countries where corruption is rife, without resorting to bribery? I have been in this business for twenty-two years. I just don't buy it. It's impossible. We need to face it and accept that the DOJ is not an independent body, and that it has long been in the grip of the most powerful American multinationals. As I delve deeper into this subject, I discover that when American majors are prosecuted (which thankfully does sometimes happen), it is nearly always upon the initiative of other countries. The United States then manages to get hold of these foreign investigations in order to deal with them 'at home' and to their satisfaction.

* See Appendix 3.

The KBR/Halliburton case is a perfect illustration of this. In the mid-1990s, the American KBR, a subsidiary of Halliburton then headed by Dick Cheney, the future vice-president of the United States, joined forces with the French Technip and the Japanese JGC and Marubeni (the same as in my Tarahan case) to supply equipment for the Bonny Island oilfield in Nigeria. To win this $2 billion market, KBR paid, in the name of the consortium, a bribe of $188 million to Nigerian leaders through a London lawyer. The scandal was leaked and landed on the desk of a French magistrate, who brought charges against the London intermediary in May 2004. The Americans had no choice but to open an investigation. Finally, France and the United States made a deal; the French magistrate decided not to pursue Halliburton and its subsidiary KBR, since the American authorities had launched their own investigations. American prosecutors then discovered that KBR's leaders had received huge kickbacks. It was impossible not to indict them. However, the fines imposed on them turned out to be extremely modest in relation to the amounts involved. For instance, Albert 'Jack' Stanley, CEO of KBR, who had arranged for the payment of this $188 million in bribes and received $10 million kickback himself, only received a thirty-month jail term. KBR was fined a total of $579 million and Technip $338 million. So in a case unearthed by a French judge, a French company was ordered to pay $338 million to the US government rather than to the French government itself. This is known as shooting yourself in the foot.

And, in comparison, I risk fifteen years in jail, even though I was a mere mid-level manager who strictly followed internal procedures at the time the Tarahan contract was concluded, because Patrick Kron decided not to cooperate with the DOJ at the start of the investigation. What's more, it was on a far smaller scale than the KBR case and I didn't make any personal gain. How can there be such a disparity in the penalties imposed? The further I get in my readings, the more frustration and disgust I feel.

I note that in the US system, and not just in my case, everything revolves around a deal. When the DOJ suspects that bribes have been paid, it promptly contacts the CEO of the company in question. It then suggests several options: either the company agrees to cooperate and incriminate itself, and a long negotiation process begins (which occurs in 99 per cent of cases), or the company resists, preferring to go to trial (out of several hundred FCPA cases I've studied, that's only happened twice), or, the company tries to outwit the DOJ by dragging its feet (as in the Alstom case), but at its own risk and peril.

As a result, the overwhelming majority of companies prefer to negotiate with the DOJ and/or the SEC and reach a deal.

Regrettably, for me, it did not happen in this way. Patrick Kron apparently tried to make the DOJ believe the company was cleaning up its act, but he was playing with fire. At that point, the FBI's massive machinery, which is endowed with incredible resources, fired up. The US authorities have, in fact, enshrined the fight against corruption as the second national priority, just after the war on terrorism and at the same level as the war on narcotics trafficking. More than 600 federal agents are working to implement it, including a special unit, the ICU (International Corruption Unit), exclusively responsible for tracking down foreign companies. The FBI does not hesitate to snare companies by setting up sting operations (a deceptive operation designed to entice a person to commit an offence), which is prohibited under French law except in the fight against narcotics trafficking. By way of example, in 2009 the Americans used the services of several undercover agents (including a Frenchman, Pascal Latour) to pose as intermediaries acting on behalf of the Gabonese Minister of Defence. These false intermediaries then solicited a good twenty companies by promising them contracts in return for kickbacks. The whole thing was of course recorded. Similarly, the Americans do not hesitate to recruit informants inside companies, as happened in my own case, in order to gather evidence. The FBI will stop at nothing to provoke their targets or break recalcitrant companies, and woe betide those who try to resist them.

The American FBI may well be one hellish machine, but the further I get in my research, and the more I read, the more bizarre my situation appears. Even considering the strategic errors made by Patrick Kron that led to my incarceration, the way they have dealt with me is quite unique. It is unlike any other FCPA case.

# 23

## My guilty plea

I had a closer look into the case of a former French executive, Christian Sapsizian, a former deputy vice-president of Alcatel for Latin America. His case, which dates back to 2008, is very similar to mine. It includes the hiring of consultants to secure a contract in Costa Rica with ICE, their local telecoms operator. Until 1998, Alcatel and Alstom belonged to the same industrial group. I joined Alstom via Alcatel Câbles, a subsidiary in which I spent sixteen months as a national service volunteer abroad in Algeria from 1990 to 1992. The internal processes for selecting consultants, until they split in 1998, were almost identical. At Alcatel, just like at Alstom, consultants were paid in several instalments. The only difference between his and my own situation, is that Christian Sapsizian, the Alcatel executive arrested in Florida, drew personal gain from the situation. He picked up $300,000 in kickbacks.

Yet, when I examine his indictment in detail, I can see that the sentence he faced was way lower than mine, just ten years in spite of admitting that he made a personal gain, compared to my theoretical 125-year sentence. When I question Stan on this point, he adroitly explains that even though the law is enforced in the same way at federal level, there may be subtle differences between Connecticut, where I am being prosecuted, and Florida, where Sapsizian was indicted.

'This is precisely the reason we have sentencing guidelines,' he continues, 'to redress such situations.'

He does not seem to think that I have been singled out, or that I would gain anything by engaging in a power struggle with the DOJ. As I feel I am getting nowhere with Stan, and unsure of the next step, I turned to my cell mates for legal advice and, notably, Jacky, the veteran of the French Connection drug ring.

Having spent so long behind bars, Jacky considers himself to be far more knowledgeable than most lawyers, and he may not be entirely wrong. For years he has been writing his legal motions himself, asking his lawyers just to review them and send them to the judge.

'You have to box in the judges and prosecutors,' he advises me. 'With a binding plea [a guilty plea with an agreed-upon sentence], you bind them. You make a deal with the prosecutor on a pre-determined sentence, then you sign, and no one can go back on it, not even the judge. I hope that's what your lawyers have negotiated with the attorney.'

'I have no idea. They told me that the prosecutors had proposed a six-month sentence, so I suspect that's the case.'

'You suspect? You must be sure. You must not sign an open plea that does not define the length of the sentence, because if you do, the prosecutors will screw you big time at your sentencing. The district attorneys are free to do as they please. You understand? It's called getting your arse fucked . . .'

I wouldn't have put it quite that way, but he may be spot on. My defence lawyers both worked as assistant attorneys before practising in a law office, so they should be aware of all the ploys used.

Why didn't they tell me there are different kinds of guilty pleas?

If Jacky hadn't mentioned it, I wouldn't have known.

The next day, I call Stan to ask if my guilty plea is binding.

'No, this is not a binding plea. That type of guilty plea is not used in Connecticut, although I acknowledge it is used in many states, including Massachusetts and New York. The folks who told you about it must have had their cases within these jurisdictions.'

'One more peculiarity of the state of Connecticut that works against me. Then what kind of guilty plea do you want me to sign?'

'An open plea.'

'So how can I be sure that my sentence will not be longer than six months if it's not stipulated on the guilty plea?'

'In Connecticut, it's the only option. Judges don't like to be pressured. We all trust each other. Judges, lawyers, prosecutors, we have been working together for decades. No one goes back on their word. If Novick tells me six months, it'll be six months. Believe me, do not worry about that. But during the plea hearing, when the judge asks you if the prosecutor made you any promise about your sentence, you have to answer "no".'

'What do you mean "No"?'

'Technically, it is not a "promise" but an "understanding" that we have with the prosecutor about the six months. But don't worry about that because we have a new problem to deal with.'

'Oh, yeah, what now?'

'You will have to plead guilty on two of the ten counts and not just one, as initially envisaged.'

'What? Yet only a month ago you assured me of the contrary?'

'This is indeed what I had discussed with Novick, but the decision was not taken by him, but by Kahn's boss in Washington, at DOJ headquarters.'

'But why did they change their minds?'

'They compared your case to Alcatel's, in which Sapsizian had to plead guilty to two counts.'

'But Sapsizian made approximately $300,000 in kickbacks, which is totally different to my situation. I get the impression, Stan, that you say yes to absolutely everything the DOJ proposes without putting up a fight. Present them with some cases where employees have not made any personal gain, please! That's your job.'

'Yes, we will, but I am afraid it won't change anything. Please remember, Frédéric, you must agree to plead guilty before Pomponi does. Otherwise your bargaining position will be undermined.'

However disproportionate and outrageous, the DOJ proposal to strike a plea deal is 'take it or leave it'. Once again, I am faced with an intractable dilemma. I have to choose between the lesser of two evils. Damned if you do and damned if you don't. I kick a few ideas around in my head, trying to decide. I even use equations: either I agree to plead guilty to two charges, with a possible ten-year sentence (which will in fact be six months if I can trust Novick), or I refuse and go to trial, taking the risk of being sentenced to between fifteen and nineteen years. I am convinced that this new blow from the prosecutors is as much a message to Alstom as it is to me. Another wake-up call to top management. 'Look what we are capable of. Look what will happen to you, if you don't fully cooperate with us.' In this whole mess, I am clearly nothing but a mere instrument, a hostage, held captive for reasons that are beyond my comprehension.

Stan and Liz, despite their apparent disappointment, inevitably prompt me to accept it. It just about kills me to do so, but I agree to plead guilty to two counts. I have no choice. But before I do, I ask them to send me the text I will have to sign.

Once again, I discover the very specific requirements of a guilty plea US style. I must pledge never to publicly contradict my admission of guilt, I cannot appeal the sentence, nor must I mention the Tarahan case when drafting my sentencing memo. My defence must only contain personal arguments (family, education, religion ...). I therefore won't be able to present my version of the facts, nor explain my position in Alstom's hierarchy. So how will the judge be able to assess my role in relation to that of other Alstom managers, whether indicted or not?

'She will be given the version of the facts presented by the prosecutor,' Stan replies.

Even more curiously, the calculation of points and the theoretical penalty range, based on the sentencing guidelines, have not been included either, which is inconsistent with all the guilty pleas I've studied. When I ask Stan about it, he replies that this is customary practice in Connecticut and that, once again, it's 'take it or leave it'.

So I finally said yes. I was summoned to New Haven on 29 July 2013 to sign my guilty plea. The same judge who three months earlier had refused my release on bail, presided over the hearing. I have now been in detention for one hundred days. It feels like an eternity.

'Frédéric Pierucci,' says presiding judge Joan G. Margolis, 'before receiving your guilty plea, I would like to ask you to take the oath. Clerk of the court, please place the accused under oath.'

The clerk then simply asks me to stand up and to raise my right hand. The hearing can begin.

'Mr Pierucci, you understand now that you have sworn on oath to tell the truth; that what you say may be liable to prosecution for perjury or misrepresentation?'

'Yes, Your Honour.'

'Can you give us your full name, your age, and tell us if you have any academic degrees?'

'Frédéric Michel Pierucci. I am forty-five years old. I graduated as an engineer in France. And I also hold an MBA from the University of Columbia in New York.'

'Do you understand the English language?'

'Yes.'

'Have you had any problems with your lawyers?'

'Your Honour, I am being held at Wyatt, so it is not straight-forward . . .'

My lawyer rises, barely giving me time to finish my sentence.

'Your Honour, given that Mr Pierucci only has limited phone calls, it's not easy to talk to him, but my colleague Liz and I did meet with him three times and we were able to talk to him without difficulty today.'

I get it. I'm just here to recite my guilty plea, rehearsed in detail with Stan. This is not the right time to be whining and certainly not the right time to criticise the American justice system.

'Mr Pierucci,' continues the president quietly, 'are you taking any kind of medication?'

'Yes. I am on tranquillizers to manage the stress of my incarceration and to help me sleep.'

'Does this medication interfere with your understanding of the debates in this hearing?'

'No, Your Honour.'

'In the past forty-eight hours, have you taken narcotics or consumed alcohol?'

'No, Your Honour.'

'Has your lawyer informed you of the maximum sentence you are facing and discussed it with you?'

'Yes, Your Honour.'

'You have therefore fully understood the agreement that you are about to consent to?'

'Yes, Your Honour.'

'Have you been subjected to any threats? Physical threats?'

How to answer this question? I haven't, but being held in a maximum-security jail without complete access to my indictment file constitutes a threat in my eyes. But if I report it, it means I can't plead guilty. So I say no and the American justice system continues to run its course.

'Mr Pierucci,' the judge emphasizes, 'to make sure you understand the scope of your decision, I would like you to summarize in a few words what you have done, and what you are guilty of.'

This is the moment I have to recite the verse I have prepared with Stan Twardy, who got it reviewed and approved by Novick beforehand. A diatribe in which I admit my guilt. I begin.

'Your Honour, between 1999 and 2006, I was Vice President Global Sales and Marketing of the Alstom Boiler business . . . At the time, I was based in Windsor, Connecticut. From 2002 to 2009, a conspiracy was engineered between different Alstom Power employees, other Alstom entities, employees of our partner Marubeni, as well as external consultants. The purpose of this conspiracy was to pay bribes to foreign officials to obtain the contract for a power plant in Tarahan, Indonesia. I, and my co-conspirators, disguised these bribes as commissions. I and my co-conspirators exchanged emails to discuss the details of this transaction. I and my co-conspirators succeeded in obtaining the Tarahan contract.'

'Thank you, Mr Pierucci. Is prosecuting attorney Novick satisfied with this statement?'

'Absolutely, Your Honour,' says Dan, who also plays his role perfectly.

'Mr Pierucci, let me summarize. You are therefore pleading guilty to two counts. Each count is punishable by five years' imprisonment and fines of up to $100,000 for the first count and $250,000 for the second. Your guilty plea may also have repercussions on the immigration laws in our country. Are you aware of this?'

'Yes, Your Honour.'

'Now listen to me carefully, Mr Pierucci. You will be questioned by a probation officer soon. This official will be responsible for drafting a pre-sentencing report. This report will then be submitted to a court to assist it in determining the appropriate sentence to be applied to you. Do you understand?'

'Yes, Your Honour.'

'This report must be submitted no later than 10 October and the prosecuting attorney will have until 17 October to respond. The court will meet on 25 October 2013 to decide your sentence. You must, of course, attend this hearing.'

'Yes, Your Honour.'

'Okay, then we are finished. I wish you all a pleasant afternoon and a good holiday.'

'Your Honour' does not joke. She is deadly serious when she

says this. The dates more or less correspond to the six months of detention that the prosecutors mentioned to Stan, so that gives me some relief. I now have a timeline for my release: 25 October.

In the meantime, on 29 July 2013, judges, prosecutors and lawyers are entering the holiday season. Connecticut is scorching hot. The armoured vehicle that brings me back to Wyatt is a furnace. Next to me, a young detainee holds his head in his hands. He has just been informed of his final sentence, ninety-six months for drug trafficking. I'm suffocating under the heat, but I comfort him as best I can.

'You can get your sentence reduced by 15 per cent for good behaviour, and get out at age thirty-five. You will have your whole life ahead of you, time to raise a family, find a job ...' I say this for myself as much as for him. *All is not lost.* However, in the heat chamber of the van, these words ring hollow. What chance does a thirty-five-year-old black male have to rebuild his life after so many years in detention? What can this country offer him? What about my future too? The van is as hot as hell and I am on the verge of passing out.

# 24

## Clara's visit

Here she is, on the other side of the bullet-proof glass. As beautiful as ever. With her long black hair and dark eyes. I have so far managed to dissuade my father from visiting me, but Clara wouldn't hear of not coming. On 5 August 2013, late afternoon, she walked in through the prison doors.

That morning, like all inmates who have visits, I shaved to make myself as presentable as possible. She must see me at my best. I patted my cheeks to give them a little colour, but my complexion is still as pasty as ever. The lack of sleep, lack of daylight, and the perpetual stress have burrowed huge bags under my eyes, and the skin on my eyelids even has a dark-purple tinge. What if she is disgusted by my appearance? I rest assured: Clara has a strong character and is very resilient. I know she's going to act as if everything is fine and smile a lot. I miss her smile more than anything.

At 7 p.m. I enter the visiting room and I finally see her. The thick glass wall separates us. I can look at her, but I cannot touch her, let alone hug her. I'd give anything to take her in my arms. However, Wyatt imposes very stringent rules on visitors and makes no exceptions. These rules contain no less than thirty-four clauses. For example, many garments are banned for women: no shorts; no dresses or skirts more than six inches above the knee; no low neck-lines; the wearing of a bra is compulsory, but it must not contain any wire; no coats; no hats; no gloves; no scarf; and no jewellery except for wedding rings. Men, for their part, are not allowed to wear hoods. It is also strictly forbidden to carry a pen or a sheet of paper. Note-taking is prohibited. All conversations are recorded.

Order and discipline supposedly prevail, but the visiting room is an absolute shambles. Imagine a vast room split in two by a glass partition. On the one side, you have the inmates (about twenty); on the other side, their families. Conversations are conducted via a telephone handset. Everyone speaks at the same time, many in Spanish. To be heard, you must press your face against the glass and shout.

Clara left Singapore to fly to France, where she just had time to drop the children off at my parents, before flying to Boston straight afterwards. She appears exhausted by this arduous journey. She stares at me, shyly, in my jailbird khaki outfit. Though she puts on a brave face, I can see that she is distraught and has tears in her eyes. In the commotion of the prison, surrounded by all these inmate families, the reality of prison catches up with her. She can no longer pretend it's not real, no longer shield herself from it as she has been trying to do for the past four months. She sees the violence flaring up, she touches the greasy walls and breathes in the distinctive prison odour. From now on, she will never forget Wyatt. Once she is satisfied I am in good health and to disguise her uneasiness, she starts talking incessantly. About the children, her work in Singapore, her colleagues, my mother, my sister. I listen to her, without saying a word. It is tremendously refreshing to hear her talk about normal life.

Conversely, when people start to discuss my situation, I get despondent. In the first few weeks following my arrest, my colleagues and notably Wouter van Wersch, the representative of the International Network in Singapore, telephoned Clara on a regular basis. Then, upon orders from headquarters, contact was abruptly severed, and Clara found herself isolated. She nonetheless asked for an appointment with Patrick Kron at headquarters. The latter delegated Philippe Cochet to meet her, the head of Alstom Power, with whom I have always had an excellent relationship. Philippe was to receive Clara on 5 August at Alstom's Paris headquarters.

We had great expectations of this meeting to prepare for the future and find out how Alstom, despite the constraints imposed by the DOJ, would help us. Unfortunately, Philippe cancelled the appointment the day before on account of my guilty plea made on 29 July. He informed Clara that henceforth he would no longer be able to communicate with her. That's it, we have now become outcasts. Clara was extremely upset, and so was I.

In addition, increasing pressure was exerted on Alstom. On 30 July, the day after I pleaded guilty, there was an 'unexpected twist' in the investigation led by the Department of Justice. I have added quotation marks to these words, as sometimes I genuinely wonder whether the whole script was not pre-written by the prosecuting attorneys, who since the very start had been diligently weaving their web, seemingly knowing each pattern in the tapestry.

Alstom's initial cooperation following my arrest did not entirely convince them. So, they decided to strike a heavy blow and indict another executive, someone more senior than me at the time of the events.

It was Lawrence Hoskins, former Alstom International Network Senior Vice-President Asia, one of the three signatories of the final approval sheet authorizing the establishment of the consultant contracts. Lawrence was hierarchically only two levels down from Patrick Kron, who himself must have felt the threat of an indictment closing in on him. According to the indictment, which was posted on the DOJ website that same day, Hoskins was accused of having known about the bribes and of having concealed the use of consultants in the Tarahan contract. If it was true that, at the highest level of management – and Hoskins was at the very top of the ladder – everyone knew about the corruption mechanisms being implemented by the international network teams, then this should surely partly get me off the hook.

It also shows that the DOJ understood the roles and responsibilities of each party.

I think I now know Dave and Dan well enough to understand some of their tactics. At first, I am particularly surprised that the DOJ publicized Hoskins' indictment, jeopardizing his arrest, whereas they went to great lengths to keep mine under seal. But since Hoskins is a British citizen and since the UK extradites its own citizens to face prosecution in foreign countries, I guess the DOJ did not take too much risk. The positive effect for the DOJ to make Hoskins' indictment public is clearly that it now puts a lot of pressure on the CEO. They are moving up the ranks in Alstom's hierarchy and have almost reached the summit. After Hoskins, next on the list should be Kron himself.

Back at the Paris headquarters, Keith Carr is probably manoeuvring hard to plan Alstom's response or rather its surrender. Naturally they will make a deal with the DOJ. It's not like they've got a choice. Alstom may well be a French major, but when faced with the FBI and DOJ, it is insignificant and will capitulate like all the others. It will be on the receiving end of a gigantic fine, but what sort of deal will my managers have to make to get themselves out of the quagmire I am already in. Who are they going to sacrifice? I dare not think about it, nor even discuss it with Clara.

The one-hour visit is over. But she will be back in two days. Until then, she will be staying with our faithful friend Linda. I don't want her to leave. This first visit was so brief. To sense her anxiety is heart-breaking for me. Yet, walking down the corridors to my cell, I feel a sense of relief. She assured me about the children and my parents. Just seeing her again has eased my burden.

Among Wyatt's prison staff there are three categories: a few nice guys, a majority who are not interested in what happens to us, and then the vermin. On the afternoon of 7 August 2013, the day of Clara's second visit, I happened to encounter a particularly obnoxious pod correction officer (CO). Although my visit was supposed to begin at 1 p.m., this CO was gossiping on the phone for ages. I was worried that if she did not stop yapping, the visitor reception office would not be able to alert her that my wife had arrived. I tried to catch her attention once, twice, thrice ... But she refused to pay attention to me. Finally, at 2 p.m., she told me that I may proceed to the visiting room. Other COs kept me hanging around in the halls again. I lost my temper, which I should have learned by now just makes things worse. As a result, I have all the henchmen on my back, screaming: 'You're in jail here, so if we want to keep you waiting three hours, that's our privilege.' It takes about twenty minutes to pass through all the doors and controls. Meanwhile, Clara has been patiently waiting for two and a half hours.

Fortunately, the next day we will have one final visit, which will exceptionally be extended to two hours. This time, despite the din of the inmates' family members, the cries of the guards trying to maintain order, the chairs scraping the floor, the doors opening and slamming, the cries and insults of prisoners who lose it, we rekindle our intimacy.

We talk about the early years of our relationship, how we first met, the difficulties we have overcome; then all of a sudden and quite incredibly, in spite of the glass partition separating us (or maybe because of it), I feel the closest I have ever felt to my wife.

Clara has left now. She will try to decompress in the three weeks of vacation she has left. The major decisions have been taken: plead guilty, enrol the children in school in Singapore for the start of the new school year, remain in her job until June 2014, when the school year ends. By then, my sentence will have been pronounced, I will have been released, and I will be able to join her hopefully before Christmas.

I have no idea what will happen next with Alstom, but the management has just appointed Tim Curran, the American head of the power business, to hold the fort in my absence. I see this as a good omen. This implies they are 'keeping my place for me'.

They're not going to fire me. I count the days during this long month of August until my release. Only two months to go.

Meanwhile, all the A pod has been transferred to the L2 pod, the one that houses the gangs. There, each individual cell of seven square yards has been converted into a double unit. This time I have been given a Greek cell mate, Yanis, with whom I fortunately get along well. But in the L2 quarter, we have no access to an open-air walk yard.

At the beginning of September, I was visited by my father who, despite my urgings, insisted on making the trip.

If I had been in his shoes, I would have done the same, so I can't blame him, but when I see him in the visiting room, I get a shock.

He is bent double, has great difficulty in walking and is using a stick. My poor father, who is normally so energetic, suddenly seems to have aged ten years. He informs me that he has been suffering for several weeks from severe sciatica that prevents him from sitting down, and that he is bedridden. It's a miracle that he has been able to cope with a seven-hour flight in economy class from Paris to Boston, rent a car and drive the three hours to Wyatt. Riddled by guilt, I start to imagine the pain and suffering I have caused my nearest and dearest. Both Clara and my father visit me three times at Wyatt, and each visit adds meaning to my pitiful existence here.

# 25

## I've been fired

An uppercut. Though, to be more accurate, it's a low, cowardly and dishonest blow.

I received the letter this morning. It is dated 20 September 2013.

<u>Subject</u>: Notice to attend a preliminary dismissal meeting.
This letter is to inform you, with regret, that we have no alternative than to terminate your employment with us. We are aware that your detention in the United States shall prevent you from attending this preliminary meeting. Therefore, please find attached the reasons on which we have based this dismissal procedure. We invite you to submit your comments in advance in writing.

I had a feeling that my decision to plead guilty would have consequences. Markus Asshoff, employment lawyer and partner at Taylor Wessing law firm in Paris, whom my family selected to defend me against Alstom and who has given and continues to give me the most loyal support, promptly alerted Clara.

In theory, Alstom had two months from the date of my guilty plea to dismiss me. I waited until the end of this period with a degree of anxiety. Somewhat naively, I was convinced that they would find a solution and halt the process. How deluded can you get! Since my arrest, they have left me to languish in this hole without even enquiring after me or offering the slightest sign of encouragement. Worse still, when several members of the executive committee travelled to the United States on business, not one of them deemed me worthy of a visit. Despicable bastards. They prefer to leave me here to rot and then coldly and calculatedly fire me!

In the document he sent me, Bruno Guillemet, the group's human resources director, starts by blaming my 'absence from work'. 'Your pre-trial detention has prevented you from executing your employment contract … Given your level of responsibility, your absence makes it difficult, indeed impossible, to maintain our contractual relationship.'

The human resources director then dwells on my guilty plea.

'Your conviction', he writes, 'will shortly lead the American judicial authorities to sentence you to a term of imprisonment, and this situation undeniably damages the image of the Alstom group worldwide. The very nature of your actions, which run counter to the values and the ethics of the Alstom group, has generated an atmosphere of suspicion and mistrust by the regulatory authorities, particularly in the day-to-day management of our business activities worldwide.'

No matter how many times I read this shameful rebuke, I can hardly digest it. Fired because I didn't turn up for work! They dare to use my incarceration as grounds! This is how they have bypassed the statute of limitations. I haven't been fired because of the Tarahan case or my guilty plea, but because I didn't turn up for work at my Singapore office. Like I had a choice. They also blatantly reproach me for pleading guilty when they know only too well that I had no alternative. This letter constitutes the height of hypocrisy. Does the human resources director really understand the force of his words? When Patrick Kron himself, bowing under pressure from the DOJ, is finally forced to admit the culpability of the company he has been running for over ten years and thus admit his own guilt, will he also be dismissed, as well as all members of the executive committee, starting with the human resources director? I doubt it. So how can they attribute acts to me that 'run counter to the values and the ethics of the Alstom group'?

Need I remind them that during the twenty-one years I spent at the company, I applied to the letter the processes predefined by Alstom's management?

Equally monstrous are their allegations that I 'violated my duties of probity, honesty and loyalty'.

Was I the one who opted to use consultants? Was I the one who chose to funnel the worldwide contracts of our consultants through our Swiss subsidiary, to conceal them? Was I the one who decided to pay bribes? And was I the one who set up the international network, the compliance structure, the cosmetic procedures for selecting consultants, and so forth? No, I strictly observed all instructions, like any manager with functions similar to mine. Besides, for the past ten years, the group, or its subsidiaries, has been prosecuted, convicted or suspected of corruption in about ten countries: Mexico, Brazil, India, Tunisia. Also, Italy, Great Britain, Switzerland, Poland, Lithuania, Hungary, even Latvia. I could go on ... Two Alstom entities were also caught red-handed by the World Bank, which blacklisted them in 2012 as part of a corruption case involving a hydroelectric dam in Zambia. And Alstom dares to claim that I have undermined the company's reputation, even though I was not involved in any of these contracts for which the group was prosecuted or convicted. This is absolutely outrageous.

The head of compliance at the time of the Tarahan events, Bruno Kaelin, also legal representative of 'Alstom Prom', the Swiss company that handled most of consultant contracts, was even arrested by the Swiss police in 2008 and spent some forty days in jail. Alstom agreed to pay a fine of tens of millions of euros to the Swiss authorities in 2011 to halt the proceedings.

The truth is stark. Alstom implemented and maintained a global bribery scheme that prosecutors described as 'astounding in its breadth, its brazenness and its worldwide consequences'.

This is something that top management knows better than anyone.

I therefore consider it rich of them to wait till I plead guilty, lecture me on integrity, and then say I have tarnished the company's reputation. They have been caught in the act, ambushed by the Americans, who possess much more powerful means of reprisal than the World Bank or the prosecution authorities in Latvia or Switzerland.

Then, at Paris, they finally cave in. After refusing to cooperate with the FBI for three years, Alstom's management strives to provide evidence of their 'good faith' to the DOJ. They are determined to show the Americans that they are now prepared to make sacrifices.

And I am the one they decide to sacrifice.

When Clara received a copy of my dismissal letter, she decided to call Patrick Kron directly. After promising her a meeting, the CEO cancelled at the last minute. Clara wrote him a letter and sent me a copy.

She evoked the conditions of my detention, which have worsened in recent weeks:

> Frédéric's physical and mental health is deteriorating every day. He witnesses things he never imagined he would live to see, including the rape of an elderly prisoner in a nearby cell, an attempted murder by concealed glass in food, the suicide of a prisoner in an adjacent cell, and the death of a prisoner due to lack of medical care, as well as recurrent fights between fellow inmates using knives.

Clara rightly complains about Alstom's lack of support.

> Frédéric is currently being detained 9,000 miles from our family home. Our children are in a state of permanent emotional distress that cannot be pacified. Our youngest twins, Gabriella and Raphaella, aged seven, cry every day for their father.

My wife explains that the dismissal letter I received 'adds insult to injury' in view of the pain she is experiencing. She reminds Patrick Kron of my long-standing allegiance to the company, that I never dissimulated or kept anything secret from Alstom, that I always respected the hierarchical validation process and that I was repeatedly commended on my work performance, as demonstrated by 'the payment of a 100 per cent bonus' just one week before my arrest. She ends the letter by appealing to him to suspend the dismissal procedure.

Patrick Kron did reply. He expressed his 'sympathy for the difficulties encountered by my family', referring to me informally as Fred, going so far as to maintain that 'this situation greatly affects him personally' – but then he actually uses the same arguments put forward by his human resources director. According to him, I have 'admitted violating Alstom's internal rules of procedure and its ethical values'. This, of course, is not true: I have never violated these rules; quite the contrary, all I did was apply them. As the company's CEO, he continues, it is 'his responsibility to protect the interests of Alstom, its shareholders, and all its employees'. He then ends by asking Clara not to write to him directly, as his lawyers have advised him to avoid any contact with my family.

This is how Patrick Kron decides to protect Alstom's interests?

Let him see it through, let him schedule an appointment with the DOJ prosecutors and admit Alstom's misconduct, and acknowledge that this entire scheme was contrived to conceal the payment of bribes by disguising them as consultant contracts for the Tarahan project and numerous other cases. Let him admit Alstom's liability, and offer to resign. This would be the best proof of his cooperation and would probably get Alstom a lower fine as well as help the company to find its way out of the abyss. Would he sacrifice his own skin? No. Rather than put his own career on the line, Patrick Kron cowardly chose to pin it on one of his subordinates.

# 26

## The six months go by

'Dad, when are you coming home?'

I have avoided answering this question up to now by telling Gabriella and Raphaella that I couldn't give them a date. My 'work' in the United States was taking longer than I expected . . . However, as we enter October, I feel confident enough to tell them that I will soon be back and that we will be able to spend Christmas together. I'm heavily misguided.

As it happens, Bill Pomponi is holding out. He refuses to plead guilty. As he is an American citizen, the judges have released him, giving him time and space to prepare his defence. If what the prosecutors say is true, after my guilty plea Pomponi has little or nothing left to negotiate. He therefore risks a hefty prison sentence of at least ten years. For an elderly person in poor physical health, this almost equates to a death sentence. He therefore has a vested interest in delaying the procedure. I can understand this.

At the same time, his legal strategy will have a major impact on my fate. As long as Pomponi resists them, the prosecutors won't let me be sentenced. The reason being that if Pomponi goes to trial, they will want me to testify against him. They must therefore keep me within their sight and not let me return to France. I am again faced with this fiendish mechanism, despite trying to seek an alternative with my lawyer.

'And if I refuse? I still have the right to be sentenced within three months of my guilty plea, don't I?'

'Absolutely, it's up to you. But if you persist in pinning them to this date, the prosecution will shoot you down in flames at your sentencing hearing and request ten years, instead of six months.'

'So what should we do? We ask for my release on bail, I return to Singapore, and await a sentencing date that suits the prosecutors?'

'They will not let you return to Singapore. You will have to be kept on parole in the United States.'

This is an almighty blow. It could go on for months. My fate depends on what Pomponi decides. I must contain my fury as I'm powerless. I must accept that my hearing will be deferred. When I break the news to Clara, she is devastated, but at least the family can come to the United States for Christmas and we will have a fortnight together. She starts searching for an apartment for me to stay in once I am released on bail.

Two days later, I get a visit from Stan at Wyatt. I immediately sense his irritation.

'I have some very bad news. Not only are the prosecutors postponing your sentencing hearing but they now also object to your bail application.'

'What? It's been exactly six months. That's the agreement you made with Novick.'

'I am furious myself. We don't do that here in Connecticut, where this type of verbal agreement normally underlies the relationship of trust between lawyers and prosecutors.'

'But Stan, I don't care about what's done or not done in Connecticut.'

'Novick told me that the orders had been issued from Washington, from Kahn.'

'Yes, but you knew that from the outset.'

'I agree and I apologize. This is the first time I have encountered such a situation.'

'What do they want?'

'They now want you to stay locked up at Wyatt between six and ten months.'

'Why between six and ten months? What's behind the figure of ten months?'

'I have no idea. Neither Novick nor Kahn, whom I of course phoned, wanted to elaborate. But something is brewing.'

'What do you think it is?'

'It can only be linked to Alstom, but I don't know what.'

I have been tricked by them once again, just as I was starting to see light at the end of the tunnel. This is a huge disappointment and is extremely hard to bear, especially for Clara and the children. I cling to the hope that I could be released any time between these six and ten months, though mentally I must prepare myself for the worst. Another four months in this hell hole of Wyatt. I frantically resume my studies of all the FCPA cases of previously indicted companies and individuals to find out how they fared. Juliette or Clara send me all the missing documents. I become obsessed by the FCPA.

I desperately try to figure out what the prosecutors' game is. I'm merely a pawn in their chess battle against Alstom. Yet I am somewhat mystified by the extreme nature of their wrath. The war they are waging against Alstom seems to surpass a mere desire to punish the company. They seem to be driven by a moral duty, an almost divine mission to stamp out corruption on a global scale.

Or . . . is it something else I am unaware of?

In late 2013, besides its legal problems, Alstom goes through a difficult period. Something I learn from reading *Le Figaro*, to which my father subscribed me. Clara also regularly sends me press clippings on the company. By the time I get to read them, they are two weeks out of date, but this is of no consequence to me because at Wyatt I have all the time in the world.

I discovered on 6 November 2013, a week late, that Patrick Kron had announced 1,300 job cuts, mainly in Europe, including around a hundred in France. This decision doesn't really surprise me. In the past year, alarm bells have been ringing; the group has been hit by the global economic slowdown. European countries have not yet emerged from the financial crisis, and growth has been less than expected in developing countries. The outcome was a 22 per cent drop in orders for Alstom compared to September 2012. Furthermore, the company failed to obtain some key contracts. Eurostar chose Siemens to manufacture its new trains and Alstom was pipped at the post by Canadian Bombardier to produce new rolling stock in the Île-de-France region. SNCF the French state-owned railway operator, deemed Alstom way too expensive. In addition, in the energy sector, sales of our gas turbines were down.

Admittedly, the fundamentals were still very good: the group enjoyed the greatest nuclear expertise in the world. It was ranked number one for the supply and maintenance of turnkey power plants and equipped approximately 25 per cent of the world's nuclear installed base. The company was also a world leader in the production of hydroelectric power.

However, though the group was a long way from experiencing such an epic crisis as it had in 2003, the situation remained worrying.

Cash flow was likely to be negative again for the third time in four years. In this context, on 16 November 2013, Patrick Kron, CEO, as published by *Le Figaro*, set forth his strategy, which was to sell off part of the rail transport division to the Russians. Alstom could sell 20 to 30 per cent of its shares in this division, with an expected gain of 2 billion euros, which would jump-start activities in the energy division. This is one of the key advantages of a company with multiple sectors of activity. During low-cycle activity in one sector, the remaining sectors level out its earnings.

In this announcement of November 2013, there is, however, one question that remains unanswered: what about the alliance announced in 2011 between the boiler business and Chinese company Shanghai Electric? While Kron had touted the merits of this future alliance to the analysts in each of his previous statements, no word of it was mentioned this time. Very odd. Why am I still interested in all these stories?

Clara's letter – like the letters my parents wrote to Patrick Kron – made no difference. On 16 November 2013, Alstom fired me with prior notice ending 30 June 2014, on real and serious grounds related to my protracted absence that caused disruption to the company and necessitated my permanent replacement (and not for serious misconduct in relation to the Tarahan project and my guilty plea). Alstom was just about willing to bear the cost of repatriating my family from Singapore to France at the end of the school year.

December arrives. It has been almost three months since I last set foot outdoors. I'm suffocating. Prison life is slowly breaking me. Now I fear my next visit more than anything else. Although I asked them repeatedly not to come to see me, my mother and sister, after my father and Clara, finally decided to make the trip. They arrive tomorrow.

# 27

## The family rallies round

My appearance frightens them. A ghost-like figure in a khaki jumpsuit. I can see their surprise.

'It's unbelievable how much weight you've lost!' were my mother's first words. 'Are you eating properly here?' she asks. 'Really, are you sure you're eating enough?'

And then she starts to cry. The shock of prison, the joy of seeing me and the exhaustion of a gruelling journey have overwhelmed her.

'We landed in Boston late afternoon,' she finally tells me, 'and after waiting almost three hours for our rental car, we drove late last night to Providence [Rhode Island's capital].'

My mother is almost seventy-six years old and suffers from Parkinson's disease. She was shocked at the overt poverty in Providence.

'It was sad; I had the impression that I had landed in Fargo, like in the Coen brothers' film, as if the city had been abandoned.'

She enquires again whether I am eating properly. Like any mother would. My sister Juliette is equally moved by my plight. However, Juliette, who is well acquainted with the French penal system by dint of her profession, immediately embarks on a comparison.

'There's no getting away from it, it's very professional, very clean here.'

I smile wryly. The visiting room is the most presentable part of Wyatt. Undeniably, families are well received. However, Juliette's satisfaction proves short-lived, once the realities of the communal visiting room hit her. She finally manages to obtain an individual meeting room for their next visit, on an 'exceptional basis' and on the grounds of my mother's state of fatigue after the long journey and her ill health. Enjoying this smidgen of privacy, I listen as she recounts how she has alerted the Ministry of Foreign Affairs, in an attempt to mobilize the French authorities into addressing my situation.

'Even as far back as April, when you were arrested, I was the one who alerted the French consulate in Boston, who knew nothing about it. The French consulate in New York forgot to tell them. Then, in May, I accompanied Dad, who had an appointment at the Ministry of Foreign Affairs, where we were received by the Head of the Sub-Directorate for the Protection of Persons, and the Head of the Consular Protection of Detainees division. They were rather distant and cold, as if your situation was not their problem.'

That particular meeting had truly stuck in my sister's throat.

'They explained to me that they are responsible for handling the cases of two thousand French prisoners around the world and that your situation was not the most pressing. We tried to make them comprehend that your situation was extremely unique and that, above and beyond your case, a leading French multinational was in the cross hairs of the DOJ. And you know what they replied? "Not at all, Madame! We do not see how the French State could be involved in this matter. The case of Frédéric Pierucci is comparable to that of a small-time entrepreneur who has come unstuck for not having paid his VAT." . . . Can you believe it, Fred?'

Juliette's anger at the memory of this interview almost lifts my spirits. And when I see the efforts my sister and mother are making to help me, it spurs me on in my fight for freedom.

But the days go by so slowly at Wyatt. Christmas is approaching, and the prosecutors have still not given any sign of a possible bail date. On 28 December my lawyer Liz asks me to get in touch with her as promptly as possible. After eight and a half months of incarceration, perhaps this is good news at last? A cruel disappointment.

'We received a phone call from Jay Darden of Patton Boggs, Alstom's lawyer, who informed us that your company has decided to stop paying us, with retroactive effect from 29 July, which was the day you pleaded guilty. So you're going to have to pay our bills from August to December yourself,' Liz announces matter-of-factly.

I am speechless. When I regain my senses, I reply: 'I'll ask my family to contact the headquarters in Paris to resolve this situation. In your opinion, who was at the origin of this instruction?'

'Alstom is perhaps desperately trying to please the DOJ, or maybe the pressure wielded by the DOJ is so intense that they have to bow. It's all the same at the end of the day.'

So here I am, locked up in a maximum-security jail nine thousand miles from my family, fired from a company I served for twenty-one years, abandoned by the French authorities who refuse to lift a finger, forced to pay monstrous legal fees, without knowing when I will be released or what my final sentence will be. No matter how hard I try to stay positive, I hit rock bottom.

Early in January 2014, a glimmer of hope returns when the vice-consul Jérôme Henry informs me of President Hollande's trip to the United States in February. Above all, he assures me that my case will be evoked during this state visit.

In his view, some members of the government were starting to question the purpose of the DOJ's pursuit of Alstom.

Personally, I have no illusions, but the consulate and my parents were very hopeful that the President of the French Republic would raise my case during his one-to-one meeting with Barack Obama. My parents even wrote to the French president:

Mr President, our son is currently in pre-trial detention in a maximum-security US detention facility. You can well imagine the distress of our family caught up in such a nightmare. You will note that the two other individuals involved in this case, two former Alstom employees [Rothschild and Pomponi] of American nationality, have not been incarcerated. In this case, it cannot be ruled out that the DOJ has chosen to indict employees, due to Alstom's poor level of cooperation in this case over the last few years. We respect the judiciary and its independence. We therefore ask you to request the American executive authorities to show mercy or grant pardon to our son, as part of the constitutional powers vested in the American President. We appeal to you, Mr President, to heed this call from helpless parents and to raise this issue with your counterparts in the bilateral meetings that will take place during your state visit.

The letter fell on deaf ears. During his visit to the United States, François Hollande, who had been briefed by the French Embassy as to my incarceration, did not ask for Barack Obama's intervention. With insight, I can see that the two heads of state had a host of other more critical issues on their agenda, such as the Syrian crisis, the spread of nuclear weapons, the fight against terrorism and climate change. Not to mention intelligence activities.

Three months earlier, in November 2013, Edward Snowden's revelations put a damper on relations between our two countries. Even if François Hollande, at the end of his interviews with the American president, played the appeasement card by declaring that 'mutual trust has been restored', the revelations about the extent of the wire-tapping programmes set up by the American National Security Agency (NSA) certainly left a dent.

The documents taken from the NSA by Edward Snowden are edifying, to say the least. They establish that over a thirty-day period, from 10 December 2012 to 8 January 2013, the United States snooped on and recorded more than 70 million pieces of French telephone data. On average, 3 million pieces of data are intercepted daily, and some specifically targeted numbers system-atically trigger the recording of conversations. Some keywords also allow the retrieval of SMS messages and their content.

I wonder about other documents revealed by WikiLeaks. For example, a note entitled 'France: Economic Developments' details how the NSA's mission is to glean information on the business practices of major French companies. American intelligence services scrutinize all markets worth more than $200 million in strategic industries: gas, oil, nuclear or electricity. In other words, the majority of business sectors in which Alstom is a key player. These disclosures are proof of the extent of US commercial espio-nage. This has been a long-standing practice within the US intel-ligence community. As early as 1970, the Foreign Intelligence Advisory Board* recommended that 'henceforth economic intel-ligence be considered a function of the national security, enjoying a priority equivalent to diplomatic, military, technological intelli-gence'. James Woolsey, CIA director from 1993 to 1995 (during Bill Clinton's presidency), acknowledged in an interview he gave to *Le Figaro* on 28 March 2000 that 'it is a fact: the United States secretly collected intelligence against European firms and I consider it is entirely justified. Our role is threefold. First, monitor companies that break UN or US sanctions. Next, track the technologies for civil and military applications. Finally, hunt down corruption in international trade.'

Over the years, the Americans have developed a double-trigger system. Upstream, the sheer force of their intelligence system allows them access to the biggest deals concluded by foreign companies. Downstream, their sophisticated and extremely well-oiled legal instrument authorizes them to prosecute in total legality

---

* The information on American economic intelligence is taken from a research report by the CF2R (Centre français de recherche sur le renseignement): 'American Racketeering and State Resignation' by Leslie Varenne and Éric Denécé.

non-compliant companies and to net billions of dollars in fines. No other country in the world has such an arsenal of weapons at its disposal. It enables them to debilitate, eliminate or even absorb their main contenders. 'No individual or entity that causes harm to our economy is above the law,' sums up Eric Holder, Attorney General of the United States, in a pithy statement. But industrial firms are not the only targets. Since the mid-2000s, and particularly since the sub-prime financial crisis, the American administration has been lining up financial institutions, one after the other, that have not complied with its embargoes. Early 2014, it snared BNP Paribas, which was prosecuted for having conducted transactions in US dollars with enemy states that the United States has black-listed; for instance, Iran, Cuba, Sudan or Libya. The BNP will soon be ordered to dismiss or sanction about thirty of its senior executives and will agree to foot a hefty bill of $8.9 billion (this banking case was bad timing for me as it politically overshadowed the Alstom case). Other French banks such as the Société Générale or the Crédit Agricole will also have to settle large monetary penalties with the US Treasury.

To this day, I have not yet grasped why our governments do not take a tougher stand against this American racketeering. What are they afraid of? To what extent are our industries going to let themselves be pillaged?* Would we accept such a dictum from another state? Why do we continue to act like willing victims? We continue to observe our own decline.

---

* Since the entry into force of the Sapin II law in 2016, the French State has been able to recover part of the fines imposed in joint prosecutions with the American authorities. In the Société Générale case, France was thus able to obtain direct payment on half of the fine i.e. $292.8 million.

# 28

## My new job

Sean is a giant one-legged dude who lives in the same pod as me. Every Monday, he attends my 'chemistry class'. I am now an 'assistant teacher'. I've been locked up in Wyatt for one year. Yes, a whole year. Even in my worst nightmare, I could never have envisaged this. Twelve months spent going round in circles in a maximum-security fortress. In my misfortune, I was at least lucky enough to get a job at the beginning of March. 'Assistant teacher'. I now have a busy schedule, giving up to three hours of tuition a day. On Mondays, I teach biology and chemistry. On Tuesdays and Thursdays I teach English. And on Wednesdays and Fridays it's mathematics.

I had a long discussion with Stan and Liz. Since Patton Boggs (who were instructed by Alstom) stopped paying their fees, they have kept a low profile. As I need all my savings to pay my bail, my counsel know that I cannot pay them. However, the lawyers' code of professional conduct obliges them to continue to represent me. Not sure they're happy about it . . . I will work out how to rectify this situation later on. For the time being, this is not my priority. The most important thing is to get out of jail. After my arrest, the hours have seemed like days, the days like weeks, and my release date like a halo of light at the very end of a tunnel that dodges me every time I reach out to it. If they had told me when I was handcuffed at the airport that I was going to languish in this endless tunnel for so long, I would have screamed out like a lunatic.

Stan and Liz confirm to me that, despite their multiple requests, DOJ prosecutors remain intransigent. Twelve months have passed, the prosecutors no longer even associate my case with the possible trial or guilty plea of Pomponi, who is still resisting them. Just what is going on? I mean, what are they waiting for in order to release me? I'm confused.

In the meantime, I continue to teach. Or rather 'assist' the real teacher, Mrs Watson. A tiny lady barely five feet tall, plump and with sparse blonde hair, the sixty-year-old, twice divorced and mother of five children, has been working in Wyatt for about fifteen years. She previously taught in a juvenile detention centre. She is bursting with enthusiasm and is a real chatterbox. I don't know how she manages to stay so devoted to her profession.

In the mathematics class, I help a twenty-eight-year-old inmate whose brain cells seem to have been permanently damaged from excessive cocaine use over the years. But he's a brave fellow; he wants to succeed and I do my utmost to help him. Yet though he has been taking Mrs Watson's classes for more than four months now, he still hasn't reached kindergarten level. He just can't master additions or subtractions. It's heart-breaking to see him counting on his fingers, hiding from the other prisoners, overwhelmed by shame.

By contrast, there are other youths who amaze me. Some, who left school at the age of twelve, are rapidly able to assimilate the algebra Rule of Three or effortlessly solve second-degree equations, which were a real challenge to be reckoned with at school. These inmates would have easily obtained a place at university if fate hadn't decided otherwise. Mrs Watson also organized 'psychology and behavioural sessions'.

She has a lot on her plate, to put it mildly. The majority of the inmates she teaches were born to single mothers who had several children from different fathers. They inevitably replicate the same pattern. Some of them have utmost contempt for women, who they divide into two categories: the *bitch* (girlfriend), on the one hand, and the *baby mama* (mother of their children), on the other.

And some of them spend their time bragging about their offspring. The more kids they have, the more virile they feel. One of them is very proud to have nineteen children on his 'scoreboard', even though he has not yet reached the age of thirty. Ironically, they all revere their mothers and spend hours making the most beautifully decorated cards for Mother's Day. It is the great celebration of the year, unlike Father's Day, which is consigned to oblivion.

When I'm not teaching, I'm learning. I continue to glean all the FCPA case law I can, and I spend my days drafting tables and charts all over the place, to try to identify trends, and then send the results of my extensive research to Stan and Liz, whom I inundate with countless pages of handwritten notes. But even though I now have a thorough knowledge of the FCPA procedure, two points still puzzle me. First, why has Alstom still not signed an agreement with the DOJ, with whom it has been cooperating for more than a year now? Second, why is it that no other Alstom employee has been prosecuted since Hoskins' indictment, while the DOJ (according to the documents shown to me during my cross-examination) has ample evidence of rampant corruption within the group?

At the time of my arrest, Prosecutor Novick had indicated his one burning desire: to work his way up the chain of command to the top. I am interested to see how Patrick Kron will extricate himself from the DOJ's trap. If he genuinely cooperates, I don't see how he can avoid jail, and if he does not cooperate sufficiently, he also risks prosecution. It's like trying to put a square peg into a round hole. He has no way out. Nor do I for that matter.

In the space of twelve months, my world has turned upside down. Formerly senior manager of a multinational company, I have been plunged into the depths of human misery and high-level crime. The staid engineer with the uneventful life has now become a school teacher for hard-core criminals.

Yesterday, one of my students, finally told me why he worked so hard in my chemistry classes.

'You get it, trafficking on the streets is too dangerous a job. I want to retrain, learn how to make amphetamines myself', probably inspired by the *Breaking Bad* TV series.

Fortunately, he is not very good at it and there is little chance that he will ever succeed. Although I let him believe otherwise, I'm not actually a chemistry geek myself. At school, I wasn't the best at sparking off combustion in test tubes. In real life, I'm also not too good at spotting smouldering bombs or impending explosions. Especially not the one that detonated on 24 April 2014.

# 29

## The announcement of 24 April

On 24 April 2014, in a fraction of a second, everything becomes clear and I finally have the answers to some of the questions that have been tormenting me for months.

Each morning, I take my breakfast in the common room while watching the morning news on CNN. This is the only time of day that the television set (reserved for whites) is tuned into a news channel.

At around 7:30 a.m., the presenter announces that the French company Alstom has agreed to sell 70 per cent of its business, its entire energy division, for around $13 billion to its rival, US giant General Electric.

'This is an operation of unprecedented magnitude. This is not your average takeover!' enthuses the CNN speaker, echoing the Bloomberg agency's scoop. 'This sale is GE's largest industrial deal ever,' he continues, before concluding that 'an agreement should be finalized in the coming days'.

I almost fall off my chair in astonishment! This sale is literally stupefying. Only a few months ago, to improve cash flow, didn't Patrick Kron plan to sell 20 per cent of Alstom Transport to the Russians and create a joint venture with the Chinese in the energy sector? And here he is selling off the energy, power and grid sectors, Alstom's flagship of French industry, to the Americans? Alstom may be going through a difficult period, but it is far from doomed. I am speechless.

Unless, of course, there are ulterior motives for this sale. Maybe Kron thinks he has found a way to escape the clutches of the prosecutors, by selling Alstom's prized power and grid division, that the Americans have been coveting for so long, to General Electric, in the hope of favourable treatment by the US Department of Justice. Having studied thousands of pages of case law, however – and despite Kron's later consistent denials that he'd attempted to negotiate his immunity* – I find it difficult to imagine that he would have risked undertaking a transaction of such magnitude, which was liable to provoke violent political reactions, without having made some sort of personal deal with the DOJ.

I suppose this is also why I was not released after six months. They need a hostage. And since the DOJ is the only institution empowered to decide whether or not to prosecute an individual, all this is perfectly legitimate, at least from the American point of view. Do the French authorities know what is behind this sale? I highly doubt it. These are my thoughts minutes after hearing the announcement on CNN.

I'm very upset by this announcement. Many thoughts run through my mind. For example, I find it hard to imagine that the government would allow such an operation to go ahead, as it would severely impact on France's energy independence.

Alstom manufactures, maintains and refurbishes all the turbo-alternators of the fifty-eight nuclear reactors installed on French territory, and the group produces in Belfort the Arabelle turbines for the European Pressurized Water Reactors (EPR) that Areva and Electricité de France (EDF) build like the one in Flamanville. Alstom therefore plays a vital role in 75 per cent of our national electricity production and possesses technology that is admired the world over. Alstom also supplied the turbo-gear for our aircraft carrier, the *Charles de Gaulle*. This means Alstom is a highly strategic undertaking for our country. To allow such an asset to slip into the hands of a foreign conglomerate would be sheer madness. There is far too much

* Matthieu Aron's interview with Patrick Kron.

at stake. I cannot believe that the bid launched by GE will go ahead; the French government would never agree to it.

Nearly four thousand miles from Wyatt, a member of the French government happens to agree with me.

'I can't believe this, this is complete and utter bullshit!' unleashes Arnaud Montebourg, the Minister for the Economy and Industrial Renewal in Manuel Valls' government, to one of his colleagues who informs him of the Bloomberg agency's news leak.*

Arnaud Montebourg is not buying it any more than I do, since he takes an active interest in what happens to the multinational. From early 2013, Alstom was at the forefront of his priorities. Disquieting information had been relayed to him. Alstom is going through a rough patch. With the ongoing global economic crisis, there is a slump in the energy market, orders for power plants have fallen to lower than expected, and though Alstom is still huge in France, it is smaller than its two main rivals, Germany's Siemens and the US's General Electric. But the government's main concern was the Bouygues group's announced withdrawal. Bouygues was Alstom's main shareholder. It was seeking to sell its stake in order to focus on telecoms and notably 4G.

Financial analysts explored the various avenues that would enable Alstom to survive this economic downturn. Arnaud Montebourg assigned this mission to Roland Berger, one of Europe's leading industrial strategy consulting firms. The agency, which was founded in Germany, is present in thirty-six countries and employs 2,400 staff. Alstom's audit mission was assigned to one of its flagship consultants: Hakim El Karaoui. A 'Normalien' (graduate of the Ecole Normale Supérieure) and previous adviser to Jean-Pierre Raffarin, former French prime minister, and to Thierry Breton at the Ministry of Industry, El Karaoui is also a close friend of Arnaud

---

* Quotation from *Alstom: A National Scandal*, by Jean Michel Quatrepoint, published by Editions Fayard in 2015.

Montebourg. Roland Berger's auditors drew up a highly contrasting report on the company. They considered that Alstom has some proven strengths, but that the company needs to forge new alliances to gain leverage. In their report, they favoured a Spanish or Polish alliance in the transport sector and suggested that the energy sector should collaborate on an ad hoc basis with Areva. At no time, however, did they advise a total or even partial sale of the group.

In February 2014, the conclusions of this study started to leak into the press, Patrick Kron was highly upset and complained to Arnaud Montebourg. 'Your HEC [elite French business school] interns are nice, but they talk too much.'*

It is no secret that Patrick Kron, an elitist liberal, good friend of Nicolas Sarkozy (he was one of the guests at the Fouquets' party in May 2007 celebrating Sarkozy's presidential election), and the socialist minister Montebourg, a proponent of state capitalism, do not see eye to eye. In fact, they genuinely dislike each other. Since the beginning of 2013, however, they were nevertheless compelled to work together. They even met six times. On each occasion, the focus of their discussions was Alstom's future. Admittedly, the state is no longer a shareholder, so there is no apparent reason why the government should interfere in the affairs of a privately-owned group. But in Arnaud Montebourg's eyes, Alstom is no ordinary private company. First, the company has relied for nearly a century on public procurement. Second, it is indebted to the state, which stepped in to save it at the time of the 2003 crisis. Third, its nuclear focus and transport activities with France's high-speed rail network, the TGV, and the Metro constitute a vital interest for France. There is also a fourth, greater political motive here. How could Arnaud Montebourg let a socialist government abandon a French multinational that was rescued in 2003 from the brink of bankruptcy by conservative Nicolas Sarkozy? He was convinced that voters would

* Interview with Mathieu Aron.

never forgive him. And therefore, through numerous interviews, he has been pressing the head of Alstom to come up with a solution for nearly a year now. And that's why he cannot believe that Patrick Kron would have betrayed his trust and gone behind his back.

On 24 April 2014, a few minutes after the Bloomberg agency had leaked news of the sale, Arnaud Montebourg immediately called Emmanuel Macron at the Élysée Palace. The Deputy Secretary General (in charge of economic affairs) expressed the same level of surprise, claiming that he hadn't known anything about the deal. I wonder whether his surprise is genuine. Later on, I was to learn that Emmanuel Macron, after his arrival at the Élysée Palace in June 2012, had also discreetly commissioned a report on Alstom's future from the American agency A.T. Kearney, to study the social ramifications of a merger between the company and other major players in the sector. To what end? What other information did he possess at that time? Was he closely monitoring the American legal proceedings? It remains a mystery to this day.*

Meanwhile, Montebourg (who is unaware that Macron has conducted research on his side) urges his staff to gather all the information and confront Patrick Kron. However, the CEO of Alstom is nowhere to be found. And for good reason. He is on a plane returning from Chicago, where he has just finalized the terms of the sale with the head of GE.

Finally, the news came from New York. Clara Gaymard, Managing Director of GE France, was on a business trip to the United States that day, and confirmed to the minister that discussions between her company and Alstom were indeed underway.

Arnaud Montebourg must bow to the evidence: Patrick Kron is in the process of selling France's crown jewels to the Americans, without even informing him.

---

* We contacted Emmanuel Macron, when he was Minister of Economy and Finance, but he did not wish to answer our questions.

Kron did it behind his back. According to press reports, the deal with GE should be consummated within the next few days. A reception room, the Gabriel Pavilion, is booked to announce the event to the Paris financial and business community. Arnaud Montebourg is seething with rage. He refuses to give in to such blackmail. Patrick Kron will have to answer to him. He therefore sends a driver to pick him up from the airport upon his return from Chicago and summons Kron to his office.

The interview is stormy. The CEO tries to argue that Alstom is not facing temporary difficulties but a structural crisis.

'The company,' he continues, 'no longer has the critical mass required to survive in a rapidly changing market. A drastic solution is therefore required: sell the energy division to replenish cash flow, to revive Alstom in the transport sector.'

Arnaud Montebourg refuses to listen. He goes on the attack.

'See this desk? You're not going to be seeing it for much longer! Where you are sitting is precisely where Philippe Varin [former CEO of Peugeot SA] lost his golden parachute. And you'll not be coming back. So enjoy your coffee; it's your last one.'

Patrick Kron lets the storm pass. Later, he told his circle that he was still somewhat surprised by the crudeness of some of Montebourg's words.

'During this meeting, within the gilt-panelled walls of the room, I was told: "You tried to fuck us from behind!"'

The remark is certainly a trifle offensive, but I must admit that I find it quite pertinent. Arnaud Montebourg had reason to be furious. Patrick Kron did a superb job of hiding the sale from him, as he likewise concealed it from his executive board, his board of directors, the head of the Power division, Philippe Cochet, in spite of the latter being the most affected by it, and even his CFO. He only informed two people within Alstom: Keith Carr, the general counsel, the same person in charge of negotiations with the DOJ, and one of his close deputies, Grégoire Poux-Guillaume, president of Alstom's Grid Sector.

This young forty-something, whose father, a former Péchiney employee, was a close friend of Kron's, was entrusted with the task of reaching out to GE in utmost secrecy. I know Poux-Guillaume well and I immediately understand why he was entrusted with the very delicate task. In 2004, shortly after his arrival at the head of Alstom, Patrick Kron appointed him head of the Environment Control Systems (ECS) business at the age of thirty, in charge of pollution-control equipment for coal-fired power plants. This equipment was most often installed downstream of the boiler, so I worked on almost all the projects with his sales teams. Then, in 2007, Poux-Guillaume left Alstom to join CVC Capital, a major global investment fund based in Luxembourg. A year later, CVC Capital joined forces with GE to try to acquire against Alstom, Areva's Grid division. Even though their attempt failed, Poux-Guillaume had established close contacts with GE's top management at that time. Finally, in 2010, after his departure from CVC, Poux-Guillaume was back in Alstom's fold, and re-established his relationship with his mentor, Patrick Kron.

I would like to know the date on which Kron's 'fledgling' informed GE that Alstom's energy division was for sale. This is something I have been wondering for a long time.

'The beginning of 2014' was the official time period given by Alstom's management. Personally, I have always been convinced that these negotiations started much earlier. I then received confirmation of this. In fact, Grégoire Poux-Guillaume initiated negotiations in August 2013, nine months before the information was finally disclosed by the Bloomberg news agency. Arnaud Montebourg had plenty to be angry about. For more than nine months, Patrick Kron had kept him and the entire French government in the dark.

This timetable (long kept secret) was critical for the negotiations. It corresponded to another chronology: that of Alstom's legal setbacks and indeed my own. By piecing together my own recollections of events at the time, and what I later learned from the indictment and other Court documents, I was able to see the bigger picture.

In the summer of 2013, a wave of panic swept through the group's top management. On 29 July 2013, I pleaded guilty, and on 30 July 2013, Lawrence Hoskins, the company's International Network Asia SVP, just two levels down hierarchically from Patrick Kron, was also indicted. Back at the Paris headquarters, this caused alarm among the group's senior executives. They feared who would be next on the DOJ's list. Will the Americans try and nab the CEO?

It was precisely at this time that Poux-Guillaume approached GE's management. It is hard to believe that this timing is sheer coincidence.

It was also during this second half of 2013 that Alstom had to start negotiating a cooperation agreement with the DOJ, requiring them to fire me and then cease paying my legal fees (hence the long delay between my guilty plea of 29 July and the announcement they would no longer pay my counsel on 28 December 2013).

After that, I also learned that a meeting was held on 9 February 2014 in Paris at the Hôtel Bristol. Five people attended. On Alstom's side, Patrick Kron and Grégoire Poux-Guillaume, and on General Electric's side, the CEO Jeff Immelt, accompanied by his mergers and acquisitions manager and his head of energy division. At that point, neither Philippe Cochet, the head of Alstom's Power division, nor Alstom's CFO, were any the wiser. This is unprecedented for a transaction worth €13 billion, but first and foremost, why were they kept in the dark?

Once again, this timeline is disquieting. The meeting in the Bristol hotel was organized at precisely the time that Alstom learned that it would have to pay a mega fine. At the time, according to the *Washington Post*, analysts at Nomura (a financial consulting firm) estimated that it could reach $1.2 billion or even $1.5 billion. Moreover, even though Patrick Kron has always fiercely denied that the corruption prosecutions influenced his business choices, I just don't accept it. In business, coincidence is rare.

And I'm not the only one who thinks this way. At the highest level of the state, Arnaud Montebourg also expressed concerns. To get a clearer picture, in April 2014 he even sought the services of the French counter-intelligence services, the DGSE. He contacted Bernard Bajolet, head of the DGSE, in person, from an encrypted telephone, but the latter refused his request. The head of the French secret service informed the Minister of Industry that its services do not traditionally operate on 'friendly' territory. The DGSE does not step on the toes of an ally as powerful as the United States.

In the spring of 2014, while an American heavyweight is in the process of seizing one of our most strategically vital enterprises, we can only deduce that our economic intelligence services faltered. The former delegate for economic intelligence to the Prime Minister, Claude Revel, acknowledged in private that there had been a 'brutal failure to respond'. Then, when her department discovered that the sale of Alstom constituted yet another episode in the economic warfare that the United States was waging against its European partners, she repeatedly tried to warn the regulatory authorities; alas, to no avail.

# 30

# A time of reckoning with Stan

In the spring of 2014, I only catch snippets of the buzz caused by the sale of Alstom in France. I am primarily concerned about the consequences of this transaction on my legal situation. I call Clara, Juliette and her husband François. They have all come to the same conclusion as me – that this is a strategy by Kron to avoid being snared by the DOJ. My lawyer in France, Markus Asshoff, and Jérôme Henry of the French Consulate in Boston have also come to the same conclusion. We pull together and brainstorm. Perhaps the Ministry of Foreign Affairs will now finally take stock of the situation?

Further correspondence was sent to the Élysée Palace, the Prime Minister and the Ministry of Foreign Affairs. I am henceforth convinced that Patrick Kron will not do anything to help me, and I hope (naively) that a diplomatic intervention will get me out of here. Obama is about to come to France to celebrate the seventieth anniversary of the Normandy landings. This could be a new opportunity after President Hollande's visit in February.

Until then, I'm determined to have a frank conversation with my two lawyers. While reading the FCPA case law (which meantime has become my bedtime reading), I recalled a detail that puzzled me about GE. When I look back at my notes, the truth leaps out at me: Alstom is the fifth company to be swallowed up by GE after being accused of corruption by the DOJ. This discovery, which I later shared with journalists, was confirmed and published in *Le Figaro* on 22 December 2014.

I sometimes even wonder whether GE passed on information relating to Alstom's practices to the DOJ. This wouldn't be the first time a judicial enquiry is triggered by a competitor. Or did General Electric simply seize the opportunity to exploit Alstom's debility, especially the very fragility of Patrick Kron who was facing prosecution?

Economic warfare is brutal. Right now, I don't care about any dirty scams. All I want is to show my two lawyers that I'm not going to be taken for a fool, and to discuss with them how to take advantage of this new situation. I know that Stan Twardy, as a former prosecutor, has maintained many links with his colleagues in the Department of Justice, starting with the first of them, Eric Holder, the United States Attorney General with whom he worked directly when he was stationed in Connecticut. I hope he will question him, or at least one of his deputies.

'Stan, you have clearly understood that Patrick Kron is handing Alstom to General Electric to avoid prosecution. You must therefore report this to the DOJ.'

'I don't think that's possible,' he replied calmly.

'Why not? You know them well.'

'That's right,' admits Liz Latif, but this is an offensive question. 'You make it sound like DOJ prosecutors are in league with GE. I hope you're not challenging the independence of the American judicial system?'

'Well yes, I am!'

During the many hours I have spent researching, I have gone over all agreements with a fine-tooth comb – all agreements that companies have entered into with the DOJ – and several have all the hallmarks of political pressure having been exerted. I carefully noted them all and I furiously churn out the list to my counsel.

'Take the example of BAE, the UK defence company suspected in 2006 of corruption involving payments to members of the Saudi royal family [including allegedly $2 billion into the Washington bank account of Saudi Prince Bandar bin Sultan who was at the time the Saudi Ambassador to the United States] in connection with the £43 billion al-Yamamah fighter plane sales. In December 2006, Tony Blair the then Prime Minister of the UK, directed the SFO to stop the investigation based on national security grounds and in order not to jeopardize the sale of new Typhoon fighter jets to Saudi Arabia. Finally the US came to the rescue of Saudi Arabia and the company finally admitted in 2010 in a parallel investigation in the US by the DOJ to "accounting violations and a conspiracy charge" even though the case had been widely publicized in the press as a bribery case.* BAE, which was fined $400 million by the DOJ, was not forced to admit acts of corruption and therefore was not automatically debarred under EU rules. Besides, no BAE managers or Saudi officials were threatened after Blair's political intervention and the settlement in the US.

'Then take the SHOTShow scandal: twenty-two CEOs of American arms corporations were facing criminal charges following a DOJ sting operation, but nothing happened. At the last moment, the proceedings were miraculously aborted. Or even the Mercator case, where the consultant used by the Mobil oil corporation prior to its purchase by Exxon paid bribes to President Nazarbayev and members of his family to secure contracts on oil and gas fields in Kazakhstan. Though blatant corruption occurred, the oil company was not convicted. It should be mentioned that this whole operation was mounted with the CIA's approval. Not to mention GE. How do you account for it, Stan, that despite warnings from internal whistle-blowers reporting allegations of corruption in Iraq and Brazil, GE has got away with it each time?† I mean, have you studied the *top*

---

* After failing to report the use of intermediaries to the US Department of Foreign Affairs.
† See Appendix 2.

*ten* of the companies most heavily penalized by the DOJ? Eight out of ten are foreign, and only two are American. No Chinese company has ever been prosecuted by the DOJ for violating the FCPA. And in the forty-five years of the FCPA's existence, the FBI has never found any evidence of corruption against US oil or defence giants, but on the other hand, in the last ten years, the FBI has nabbed Statoil, ENI, Total, etc. So yes, I am questioning the impartiality of your justice system.'

'This has nothing to do with your situation,' replied Liz Latif sharply.

She is seriously beginning to get on my nerves. I'm starting to lose my cool.

'Stop taking me for a fool, both of you. Stop telling me that your justice system is irreproachable. Please.'

'We understand,' says Stan, 'but stop getting worked up. No, our justice system is not perfect. But, after more than a year of detention, you obviously still don't get it. The judge who will examine your case does not give a damn whether a deal was struck between the DOJ and Patrick Kron. He is interested in just one thing: what the prosecutors say to him. Therefore, he doesn't care if the people at the top escape punishment.'

'You're saying the judge doesn't care that he is only convicting the subordinates.'

'That's right, he really couldn't care less, Frédéric.'

'Well, let me tell you that if the DOJ protects Alstom's top executives and convicts me, it means that your judges are a bunch of Mafiosi.'

'But you still don't want to understand! Of course, the system is unjust! But it's not like you've got a choice. The bottom line is whether you want to stay behind bars for ten years or get out.'

There has never been a time when my exchanges with Stan and Liz were more strained. I feel like I'm banging my head against a brick wall.

'Stan, I don't care about your fucked-up system; I've had enough. I will be writing you a letter to formally request that you enquire in Washington at the highest level of the DOJ as to whether Patrick Kron was given immunity by the DOJ in exchange for a deal. And if you don't want to do it, it's up to you, but I would like you to reply to me in writing. That way, I have proof of your refusal.'

Stan is white with rage. He remains silent for about thirty seconds, then finally acquiesces.

'I'll pass on your question, Frédéric. But you should know that it won't do any good. It's a nonsensical and futile move.'

Our interview ends. It lasted almost an hour. No point in carrying on, as there is too much tension and too many things left unsaid, rendering any dialogue impossible. However, we plan to speak again within a week.

Just before leaving, they have one last bombshell to drop. Hoskins, Alstom International Network Asia boss, was arrested while on his way to see his son who lives in Texas. This arrest took place on 23 April 2014, the day before Alstom and GE announced their deal, at the very time Patrick Kron was in Chicago negotiating it. I see this as an obvious reminder to Patrick Kron of what could happen to him, while in the United States.

Just one year ago, the DOJ had also arrested me the day before Keith Carr came to Washington. This sequence of coincidences is simply extraordinary.

'So, Stan, that's why they changed their minds and after six months in detention refused to release me. They knew that Alstom was negotiating with GE. They were afraid that I would spill the beans to journalists or inform the French government, weren't they?'

'Maybe,' he replied evasively.

'Now all this is out in the open and the agreement is being negotiated between Alstom and GE, you can ask them when they intend to release me on bail. They can't detain me indefinitely.

Since the FCPA was enacted in 1977, there has not been a single person that made no personal gain that was sentenced to more than one year in jail. And I've been rotting in this hole for more than twelve months.'

'I will ask,' he replied tersely.

# 31

## General Electric's tale

By selling its entire energy division to General Electric, Alstom is not selling to just any industrial group.

General Electric, as I have seen in my twenty-two-year career, is more than just a corporation. It embodies America in all its omnipotence.

GE, the sixth-largest company in the world in 2014, is present in almost all strategic sectors: electricity, gas, oil, medical equipment, aviation, transport. It also manufactures household appliances, refrigerators, ovens, stoves, dishwashers and water heaters. Until 2013, it also owned one of the three largest television channels in the United States, NBC. With GE Capital, GE held one of the world's largest financial institutions. A subsidiary that bore the full force of the subprime crisis in 2008 and which, without a massive rescue operation by the American government ($139 billion), would have collapsed and caused the parent company's bankruptcy. Like Ford, General Motors and Walmart supermarkets, GE has its rightful place in every American household and is part of the national heritage.

At its helm, in this spring of 2014, was Jeff Immelt, an influential figure in Washington. He took charge of the group thirteen years before, in 2001, just four days before the attacks on the World Trade Center. His company is his whole world. His father is a former GE employee. His wife too. He has been serving the company for nearly forty years.

A formidable negotiator, Republican at heart, Immelt is also very close to Barack Obama. In 2011, the American President even made him head of the President's Council on Jobs and Competitiveness, with a mission to 'rebuild the American economy'. The big boss gets down to work, following the same course of action each time: '*Business is business.*' To him, 'doing business is also like going to war. If you're looking for love, it's better to get yourself a dog,' he says during one of his visits to Paris.

I find out in the course of my legal readings that, in 1992, General Electric managed to avoid being charged for breaching the FCPA but was nevertheless fined $69 million by the DOJ for 'conspiracy to defraud the United States and to commit offenses against the United States' in connection with a bribery scheme related to a defence contract signed with Israel. This monetary penalty sent shock waves through the company, which in turn led GE's management to embark on a clean-up mission, adopting at the same time a strict code of professional ethics and conduct.

More than any other executive, GE vice-president Ben W. Heinemann embodies this policy. In charge of compliance until the mid-2000s, he was considered by his peers at the *American Lawyer* as one of the most innovative lawyers in the United States. Under his leadership, GE established a reputation as a 'knight in shining armour' and forged close ties with the DOJ Anti-Corruption Unit. Prosecutors seeking a career change are regularly offered a place in GE's Compliance Department. In 2014, there were about fifteen of them working there. From the year 2000 onwards, the papers I was reading suggested that General Electric also grasped that other company executives who were up to their necks in corruption scandals were ideal prey. It also seemed to me that GE was ideally placed to help those executives negotiate with the DOJ while getting its hands on their companies. As I pointed out to Stan, General Electric has already acquired four companies under these circumstances in the space of ten years.

Alstom is fifth on its scoreboard and by far the biggest. In 2004, using this same technique, GE took over the American company

InVision Technologies. This company had been accused of paying bribes to secure contracts for the construction of explosive-detection systems in various airports across China, the Philippines and Thailand. GE was then closely involved in reaching an agreement with the DOJ to conclude the proceedings without InVision's CEO being charged.*

I have also noted that in the power generation sector, virtually all of GE's international rivals have been indicted and forced to pay substantial fines. For instance, the Swiss–Swedish concern ABB in 2010 ($58 million fine); the German Siemens in 2008 ($800 million fine) with eight employees being indicted including an executive committee member; Japan's Hitachi ($19 million fine); and now Alstom. But not one single leading American power engineering contractor that uses GE equipment in its offers has ever been targeted by the DOJ for violation of the FCPA, such as Bechtel (constructor of American embassies abroad), Black and Veatch, Fluor, Stone and Webster, Sargent and Lundy or even the two US boiler makers Foster Wheeler (also heavily involved in the oil sector) and Babcock & Wilcox. Yet all these majors face fierce competition in the same international marketplace for the construction of power plants, such as gas, coal, nuclear, or wind-power plants.

My question is, just how do these US companies go about securing business for the last fifty years without once getting their hands dirty? Admittedly, these companies have the backing of US diplomacy. In 2010, for example, GE was able to sell $3 billion worth of gas turbines to the Iraqi government by mutual agreement (i.e. without a tender bid) under abnormal conditions. It was all the more outrageous, given that at the time Iraq did not have the capacity to construct power plants and as a result

---

* According to the terms of the agreements entered into on 3 December 2004 between the DOJ, General Electric and InVision Technologies.

Baghdad ended up stuck with dozens of gas turbines, not know-ing what to do with them. But to this day, no one has raised any objection.

GE also generally skilfully succeeds in positioning itself as a subcontractor to a lead turnkey contractor. The US giant supplies the gas turbines to any company that is responsible for supplying turnkey power plants. Naturally it is the lead contractor that hires and pays the consultants. On the Asian market, GE has a predilec-tion for the largest South Korean or Japanese trading companies, which have also never been bothered by the DOJ.

In the spring of 2014, GE, claiming to be above all this foul play and a true leader in the fight against corruption, also proves to be a master at communication. While its takeover bid is clearly unbal-anced, its CEO Jeff Immelt, with the full backing of Patrick Kron, boasts that his proposal is 'the best way forward for Alstom'.

Jeff Immelt projects two arguments. First, GE is familiar with the French market. It has been operating in France since the late 1960s and employs 10,000 people there. Next, Alstom and GE have forged historical ties. This is undeniable, though this relationship is far from being as idyllic as Immelt would have us believe. Like many Alstom employees, I still remember the Belfort story. After licensing its gas turbines to us, GE rendered its licencees' technol-ogy obsolete by not transferring the new-generation models to them. Having rendered their licensee uncompetitive, GE then offered to buy them back.

In 1999, Alstom sold to GE its gas turbine production site in Belfort. In parallel Alstom bought ABB power generation activities which had its own gas turbine technology. This was the cause of Alstom's financial trouble in 2002/03 as ABB's technology was not mature and Alstom ended up paying huge damages to its clients who had bought the troubled gas turbines.

Be that as it may, given that GE has been present in France for several decades, it is familiar with our country, our economic structure, our thirty-five-hour working week, our culture, and above all our political networks. Its leaders are also skilled lobbyists.

In 2006, Jeff Immelt recruited the well-connected Clara Gaymard, a charming and talented ambassador who happened to be also the wife of President Sarkozy's former Minister of Economy, Finance and Industry, Herve Gaymard. She was appointed president of GE France and then promoted in 2009 to vice-president of GE International. She knows all the right people in Paris. President of the Women's Forum and named by *Forbes* magazine as the thirtieth most influential woman in the world in 2011, this public figure and *énarque* (graduate of the French *grande école* ENA) is just as at ease in ministerial offices as in TV studios.

Right now, in spring 2014, she hastens to defuse the crisis that has just broken out between her company and the French government with diplomacy, as Arnaud Montebourg's wrath does not wane. After speaking his mind to Patrick Kron in his office at the Ministry, he reiterates his arguments three days later on 29 April 2014 in the Assemblée Nationale (French parliament). The former lawyer of the Paris Bar transforms himself into a prosecutor and delivers a powerful closing speech.

'Since February,' he maintains, 'I have been questioning Patrick Kron, Chairman and CEO of this company, which is our national pride and joy, and each time I questioned him duly, solemnly and seriously, he each time replied that he had no intention of entering into an alliance. And to wrap up, must I, as Minister of the Economy and Industry, install a lie detector in his office because he isn't public spirited enough to alert the government?'

A little earlier that morning, Arnaud Montebourg, speaking on the RTL TV channel, also appealed for greater economic patriotism.

'When a deal is being closed and you forget to call the Minister of the Economy to inform him, although you normally call him for assistance on a daily basis, this constitutes a breach of our national code of ethics.'

Minister Montebourg is not content to simply employ strong rhetoric; he swings into action. He starts by courteously but firmly spurning Jeff Immelt, who had just arrived in France to finalize the agreement. Given the circumstances, he simply cannot meet him. Instead, he sends him a letter to remind him that in France: 'planned takeovers in the energy sector, and especially in the nuclear domain, are subject to the approval of the authorities'. He also issues a warning to Alstom's directors. 'Be careful,' he tells them, 'this sale could result in possible breaches of stock exchange regulations.'

In reality, this is just a show. Arnaud Montebourg is trying to buy time. He needs more time in order to organize a double counter-offensive (industrial and legal). From a judicial point of view, he is convinced that the Americans are blackmailing Alstom. However, there are several tangible elements missing from his report, which he must get his hands on before he can submit it to François Hollande.

So he seeks help from the DGSE (the French equivalent of MI6 or the CIA). The French counter-intelligence service refuses to help him. It doesn't matter; he'll dispense with their services, and get the information himself. He immediately sets up a full task force within his ministry. He appoints four advisers, entrusting them with the task of digging deep into the GE/Alstom alliance. Soon after, Montebourg's 'Four Musketeers' uncover the details of the American proceedings and just as swiftly trace it all the way back to my own case. At that time, they even tried to establish contact with me.

One of the members of Arnaud Montebourg's team calls Clara directly in Singapore. My wife is taken by surprise. Since my arrest one year ago, the French authorities have not responded. However, Clara is very suspicious; the minister's adviser appeared very young to her, and for a moment she even suspected that he might be an impostor. To prove his bona fide intentions, she asks him to send her a mail from the Ministry website. He sends an email as requested, but he is not very convincing. As for me, I am not sure what to do. All my telephone conversations and visits are recorded and sent to the prosecutors. The only opportunity I get to have a free and unsupervised discussion is when I meet my counsel. But since I dare not place all my trust in anyone involved in the US justice system, I find it very difficult to have a meaningful discussion with them.

A stroke of luck. Markus Asshoff, who is defending me in my case before the French industrial tribunal, is spending a week in the United States and has travelled to Rhode Island to visit me in Wyatt. His visit is a tremendous boost to me because, for the first time in over a year, I am able to speak to someone frankly and directly, without fear of being listened to. He stays with me for more than six hours. He has been well briefed by my sister Juliette and I can at last make sense of the many unspoken codes and innu-endoes in the telephone conversations with my family and friends. Regarding Montebourg's adviser, I cannot make up my mind what to do. Of course, deep down I want to work with him to expose the truth. But my whole family and my lawyer Markus Asshoff advise me against it.

I have been locked up in a maximum-security prison for one year now, exploited and manipulated by the DOJ and held hostage to force Alstom to cooperate, and I risk a ten-year sentence. Should the DOJ discover (and it will, because Markus suspects that he and my family are being wiretapped) that I am indirectly aiding and abetting Montebourg to thwart GE's plans, there is a high risk I could spend a lot more time here.

Reluctantly, I ask Clara to not respond to the call from the minister's office. Arnaud Montebourg continues to launch his main counter-offensive. To rebuff GE's offer, he turns to Alstom's other key rival: Siemens. The German industrial giant immediately responds with a counter-offer.

By virtue of a letter of intent sent to the French Finance Ministry, Joe Kaeser, CEO of Siemens, proposed that they take over Alstom's energy business and in return sell a significant part of Siemens' railway division to Alstom. In the shopping cart are its high-speed ICE trains (with an order book worth €5.4 billion) and its locomotives. According to the German CEO, his offer is 'a unique opportunity to create two European giants: a French transport giant and a German energy giant'. Joe Kaeser also undertakes not to slash any jobs for three years, and says Siemens is willing to relinquish Alstom's nuclear activities in order to 'safeguard France's national interests'. On the strength of this offer, Arnaud Montebourg was able to persuade Alstom's Board of Directors to defer its decision to sell to GE. The first battle was won. Patrick Kron, who wanted the deal wrapped up in three days, is going to have to revise his strategy.

However, Montebourg is losing ground. The President of the Republic has taken ownership of the case. He urgently convenes a select committee with Manuel Valls and the ministers concerned by the case. François Hollande wants to take his time too. He is wary of his overly vocal minister, who has the unfortunate habit of alienating big business. The remarks made by Montebourg about the Indian company Mittal have stuck in his throat.

'We no longer want Mittal on our territory', said Montebourg angrily in November 2012. These verbal projections pleased the left wing of the socialist party, but they annoyed those high up.

The Élysée therefore decided to appoint a mediator, in the person of David Azéma, the then head of the Agence des participations de l'État (the agency responsible for managing state shareholdings). Though the state is no longer a shareholder of Alstom, this is irrelevant, since France's national security is now at stake. Azéma is the perfect man for the job. He is a senior civil servant, with leftish tendencies but a penchant for the private sector. He reports to Montebourg but is also accountable to Emmanuel Macron.

Between the end of April and beginning of May, I try to follow this politically action-packed saga from the confines of Wyatt on CNN breakfast-time news, though the Americans are not as interested in this affair as the French. I therefore have to wait for the press clippings that Clara sends me on an almost daily basis.

Early May, I call Liz to find out whether Stan kept his promise and did in fact contact the DOJ. She indicated that he verbally asked one of his former contacts at the prosecution department of the judiciary. According to this source, she tells me no deal has been signed between the DOJ and Patrick Kron. Of course, my lawyers did not provide me with any record of this 'informal' conversation.

Since Hoskins was indicted in July 2013, almost ten months ago, it is now apparent that the Americans have abruptly halted their investigations at a certain pecking order and are no longer interested in moving up the ranks to Kron. This is a fact. However, my lawyers will say I am reading too much into it. I therefore ask Liz once more to write to the prosecutors and have them confirm in a letter that no transaction of any kind has been concluded with Kron. I feel that the prosecutors have lied to me so much from the outset that I would like material proof.

'Frankly, I advise against it,' insists my lawyer. 'I believe that the prosecutors are now ready to consider your release. They have just agreed to Hoskins' release.'

'Good for him. But you know I'm still surprised at the difference in treatment between us.'

'He is English, and as England extradites its nationals, his lawyers were able to convince the judge.'

I ask Liz: 'What are the terms of Hoskins' bail?'

'One and a half million dollars, which represents a significant portion of the house he owns in England. And he can stay with his son who lives in Texas. But if he wants to leave the United States, he will have to ask the judge's permission.'

'One and a half million dollars! That's big money.'

'It's the price of freedom. Moreover, you should be aware that you will have to pay a comparable bail package.'

'What! One and a half million dollars! But why such a large sum?'

From the start, we'd said $400,000, plus Linda's house.

'Yes, but they want the amount to be the same as that set for Hoskins.'

'It's just absurd. Hoskins may be in a position to put up that kind of money, but I am not.'

'I know, but that's how it is. In addition, you also need to find a second person in the United States who agrees to put up their home as collateral, like your friend Linda. Finally, please be aware that the two American citizens who decide to put up bail bond for you will be held jointly and severally liable, meaning that should you decide to flee to France, both their homes will automatically be seized.'

This mechanism is grotesque. The prosecutors do not stop raising the stakes. They are obviously willing to deploy any means to keep me in detention. At this point I become despondent, believing everything is over. I will never be able to satisfy their conditions and am doomed to spend the rest of my life behind bars at Wyatt.

Notwithstanding the extraordinary resourcefulness demonstrated by my family and friends in this time of need, my father managed to persuade one of his former business partners, Michael and his wife, to agree to put up their house as collateral, as Linda had previously done. I am eternally grateful to them and will never be able to thank them enough. Also, Clara, by scraping together every penny the family has, selling savings and retirement funds and even a part of the land our house is built on, managed to raise the required amount. We can't do more.

Will it suffice? I fear that my release is predominantly dependent upon the poker match currently being played out by Alstom,

GE and the French government. Yet in Paris the sale has not yet been closed and Montebourg appears to be gaining ground this time.

On 15 May 2014, he succeeded in pushing through a bespoke decree to thwart GE's plans. It was an anti-takeover measure designed to deter takeovers in sensitive and strategic sectors. Under this new law, a foreign group wishing to acquire a French company in the energy, water, transport, telecoms or health sectors would have to obtain state approval before it could be granted such authorization. Arnaud Montebourg was beaming.

'This stops us being pushed around. France must defend itself against the unwanted carving up of our industries.'

This surge of economic patriotism appealed to the French. According to a recent benchmark survey, 70 per cent of them voted in favour of the Minister's action. Will he against all odds succeed in getting Patrick Kron and the Americans to back down?

Meanwhile, throughout all this political unrest, I noticed something rather disturbing: Nicolas Sarkozy's silence and that of his party, the centrist-right UMP. How can it be that Alstom's great saviour in 2003 is not condemning the left-wing President's apathy to such a political scandal?

He could have seized this golden opportunity, but instead he chooses not to make any statement and remains silent throughout. All sorts of ideas swirl around in my mind that might explain his reluctance to become involved in the matter, such as protecting some people within his own party who are very close to GE and who will benefit from the deal.

In any case, his failure to respond is incomprehensible. I also notice that the press is sluggish. While it covers the shift in power between Montebourg and Kron, it keeps a low profile when it comes to vexatious topics; except for a few, however. For instance, on 27 May 2014, the Médiapart website, under the aegis of Martine Orange and Fabrice Arfi, published a scathing enquiry entitled:

'The ugly face of corruption* lurking behind the sale of Alstom'. In their article, the two journalists considered that 'the US lawsuits have a direct bearing on the rapidity and lack of transparency of the butchering of Alstom'. Like me, they question the disturbing calendar matches and stop at one particular date: 23 April 2014, the date at which Lawrence Hoskins was arrested in the US Virgin Islands, while, in parallel, Patrick Kron and Jeff Immelt were still in the midst of negotiations in Chicago. According to Martine Orange and Fabrice Arfi, 'this arrest was significant'. It could have been a last attempt to exert pressure on Alstom's CEO, just before he closed the deal with GE.

At long last, some of the truth is emerging. At the time I naively thought this article would make a lot of commotion. It was a flop. Like other press articles, such as the one published in the French satirical journal *Le Canard enchaîné* in May 2014, which points out what is surely a flagrant conflict of interests. In its dealings with GE, Alstom is advised on the legal side by Hogan Lovells, a US law firm. They had announced in December 2013 the nomination as new CEO of their long-time partner Steve Immelt, a specialist in FCPA investigation and ... Jeff Immelt's own brother. In the 15 May 2014 edition of the *Point*, the authors ask the only important question: 'Is Patrick Kron willing to sell Alstom to the Americans to extricate himself from a legal tight spot?', but this does not evoke any reaction.

---

* Fabrice Arfi, one of the authors of Médiapart's article, contacted Clara, but she did not pursue this for fear of exacerbating my situation.

# 32

## A Pyrrhic victory

One month later, at the beginning of June 2014, the die is cast.

Arnaud Montebourg has been defeated. In front of the cameras, he puts on a broad winning smile, maintaining to anyone who wants to listen that he is Alstom's saviour, and presents the agreement he was able to get as being the best possible deal for Alstom. Yet I am not fooled.

This is mid-battle defeat. It was not the Minister of Economy and Industry who had the last word, but François Hollande. The French President okayed the US giant's offer.

I must admit that GE, throughout the entire phase of the negotiations, worked relentlessly to achieve this end. The US titan displayed remarkable skill. Jeff Immelt, fully aware that this takeover was crucial to his career, had no qualms about taking up residence in Paris. He quickly realized that he had to win a political and media battle, both in terms of industry and finance. The CEO of GE thus teamed up with the best communications agencies in the marketplace. He chose Havas, whose deputy chairman Stéphane Fouks was a close friend of prime minister Manuel Valls.

Havas engaged very substantial resources to help GE secure Alstom. It mobilized three of its senior advisers: Anton Molina, former Deputy Director of Medef, Stéphanie Elbaz, former Director at Publicis Consultants; and Michel Bettan, former cabinet director of former Minister of Labour Xavier Bertrand. Patrick Kron, for his part, was assisted by two well-known experts: Franck Louvrier, Nicolas Sarkozy's former communication guru, and

Maurice Lévy from Publicis (very close to Clara Gaymard, president of GE France). This 'dream team' of public relations experts worked day and night to break down, one by one, the barriers blocking the sale. The first obstacle to surmount was to convince public opinion of the need for this sale.

Despite Jeff Immelt or Patrick Kron's discourse, Alstom is not a lame duck. It possesses a two-year order backlog, far more technological assets than structural defects and its current difficulties are essentially cyclical in nature. Difficult therefore to persuade the French of the necessity to divest 70 per cent of the group. Both CEOs are therefore responsible for conveying the right message to the mainstream media. Jeff Immelt will host the 8 p.m. news programme on France 2 and Patrick Kron will host the TF1 newscast. Every time Kron appears in the media, he hammers home the same old argument that Alstom does not have the 'critical mass' necessary to survive this economic downturn, especially when confronted with GE and Siemens.

But I find his arguments completely unconvincing. When you look at the figures in detail, a different picture emerges: Alstom's energy division (€15 billion) has no critical size problem; it is ranked as number three in the world energy sector. If we now compare the two companies overall, of course Alstom has a turnover eight times lower than that of GE, but the perameters are not the same and once its energy division has been sold off things will only get worse. If Alstom focuses solely on transport, it will become thirty times smaller. It is utter folly to argue that you should sell because you are not strong enough, knowing full well you will become weaker after the sale.

In addition, Kron, after saying for ten years that Alstom needed to be present in three sectors simultaneously (power, grid and rail transport) in order to absorb cyclical market fluctuations, is now advocating the exact opposite. According to him, if the company refocuses exclusively on transport it will have a very bright future.

As all specialists well know, the new Alstom is now of a subcritical size, vulnerable and at the mercy of its competitors, which, as we saw three years later, indeed provoked a takeover bid. (In September 2017, German competitor Siemens launched a friendly takeover bid on Alstom Transport. This operation was blocked by the European Commission in February 2019 despite the support, again, of the French government.) However, the message captivates the audience and hits home. It is flashed around in numerous articles and quoted repeatedly in interviews until it becomes media reality.

The second obstacle that GE had to eliminate in order to obtain the government's approval was that of employment. This was a key component in François Hollande's view. Since his election as President of the Republic, François Hollande had to contend with an unprecedented rise in unemployment. He could not risk the wrath of the labour force, by authorizing an operation that would result in a mass shedding of jobs. Immelt saved the day by swiftly making a public promise to create a thousand jobs in France. A commitment he failed to honour.

Last but not least, to convince the authorities, GE, with the support of its communications team, had to cross a third and final hurdle, undoubtedly the trickiest. They had to find a way of gagging Arnaud Montebourg.

Come mid-May 2014, the Minister of Economy and Industry made no secret of his preference for a European solution. Especially since Siemens had finetuned and upgraded its offer. It had even approached another player in the energy sector, Mitsubishi. As a result, a brand-new proposal was put on the table: Siemens and Mitsubishi. Though the latter did not want to purchase Alstom, they suggested forming a durable industrial alliance.

Mitsubishi was committed to creating three joint ventures with Alstom in the hydraulic, transmission and nuclear sectors. Alstom would remain the majority shareholder, retaining a 60 per cent stake and Mitubishi 40 per cent. Siemens would buy Alstom's gas division, but in return would sell its rail-signalling business to Alstom. Arnaud Montebourg was a fierce proponent of this solution. In his opinion it killed two birds with one stone. It enabled France to save face and was economically viable.

To stay in the race, GE realized it must revise its strategy ASAP. Taking inspiration from the German–Japanese solution, GE put forward a new idea. This time, the US giant dropped the words 'sale' and 'purchase' from its vocabulary. It henceforth proposed setting up three 'joint ventures' in nuclear power, renewable energies and transmission networks. These three joint ventures would be held 50/50 by Alstom and GE.

Immediately, the GE communications team had a documentary made to showcase the advantages of this new alliance. A fine setting at the Belfort site, where Alstom and GE employees already work together and where they can be seen eating together in a group canteen. The clip is shown at prime time on all television channels. Meanwhile, amid backroom tussles, GE's advisers are taking great pains to discredit the German/Japanese proposal, claiming it is too complex, too difficult to implement, too many partners involved, etc. Over the coming weeks, their efforts at undermining and influencing opinion finally pay off. The state negotiator, David Azéma, also sided with the Americans, but the endgame is played out at the Élysée Palace.

Early June 2014, Emmanuel Macron, Manuel Valls and Arnaud Montebourg met with the President. The Minister of the Economy and Industry argued heavily in favour of the Siemens/Mitsubishi solution, and appealed to the Head of State to use the government's new weapon, namely the anti-takeover decree he had pushed through to block GE's offer.

Emmanuel Macron took the floor.

'With Siemens, we are adding to our difficulties, and the social impact will be more brutal. Moreover, Alstom's management remains fiercely opposed to this decision.' Emmanuel Macron then goes in for the kill.

'We cannot impose terms on a private company, we are not in Venezuela.'

In contrast to the United States, which rescued GE in 2008 following the sub-prime crisis, the French socialist government, since going down the neoliberal path, is allowing a key strategic enterprise to fall into the hands of the United States. Amen. Alstom will become American. Throughout the weeks of negotiations that followed, Emmanuel Macron dissociated himself from Arnaud Montebourg.

How will the latter react to such a rejection? Will he accept it and remain quiet? Manuel Valls knows that he needs to make some concessions to retain his Minister of Economy and Industry, who has the support of the left wing of the Socialist Party within his government. Prime Minister Valls therefore suggests that the state should invest in Alstom's capital. The state could buy back Bouygues' 30 per cent shareholding and thus safeguard the company's long-term viability in the transport sector. Arnaud Montebourg has not lost face, far from it. He was able to maintain that thanks to his perseverance, the government did not abandon Alstom and that he was able to obtain some significant concessions from GE, and even argue a few days later that 'the state's investment in Alstom's capital will help guarantee an enduring alliance with GE'.

Montebourg may have done his best to conceal it, but he had been truly crushed by GE. It should be acknowledged, however, that he was the only one to defend strategic French interests.

Was he ever in the running? GE's victory owed nothing to circumstances. It reflected the omnipotence of American corporate interests on French soil. As I have witnessed first-hand throughout my professional career, the United States has a tremendous amount of influence over part of the French administration, the French economy and political class. Our leaders, even the lefties, are far more 'Atlantist' than European in their leanings. America continues to captivate and inspire, more and more. Americans are world champions in 'soft power', the soft diplomacy they use to influence and charm their partners. For example, every year since 1945, the US Embassy in Paris identifies budding French elites and invites them to Washington for a few weeks as part of a programme known as the 'young leaders'. This programme targets young political elites or *énarques* (from the ENA elite *grande école*). François Hollande, Nicolas Sarkozy, Alain Juppé, Marisol Touraine, Pierre Moscovici and Emmanuel Macron were all 'young leaders'.

US clout does not stop there, however. Today, most of the major law firms, audit firms and investment banks in the Paris financial centre are American. The Alstom/GE deal was an extraordinary boost for them, adding several hundred million euros onto their bills for services performed. To ensure effective lobbying, these institutions draw on their resources from ex-ministerial cabinet officials. The lucky elected ex-public officials can see their salary increase tenfold after departing the French administration. And never mind the obvious risks of conflict of interests. Even more shocking was the departure of David Azéma, chief negotiator in charge of managing state shareholdings in the Alstom case, to a major American investment bank. He was recruited by this bank in July 2014 only a few days after François Hollande had submitted his adjudication in the Alstom file. The former state negotiator didn't choose any old bank either. He joined the Bank of America. The same bank that advised Alstom throughout the entire negotiation.

This time, even the Ethics Committee reporting to the Ministry of Public Service found it a bitter pill to swallow. It advised Azéma to change careers. No sooner said than done, the senior official wound up in supposedly another financial institution: Merrill Lynch in London. As it happened, the Bank of America and Merrill Lynch had merged since 2008 so this change of 'official employer's name' was purely cosmetic. David Azéma then went on to serve the Bank of America/Merrill Lynch London and he had no qualms whatsoever. When asked by the newspaper *Le Monde* why he left, he replied: 'Why am I leaving the state? To earn more money.'

Another shocking move was that of Hugh Bailey who acted during this takeover period of Alstom by GE as the 'advisor on industrial affairs' in the Cabinet of Emmanuel Macron the then 'Minister of economy, industry and digital affairs'. He then joined GE in November 2017 as 'Government Affairs director' i.e. lobbyist in chief, before becoming in May 2019 General Manager of GE France!

In the war waged in the spring of 2014 over Alstom's takeover, there was one final factor to take into account. Siemens seemed to be hesitating mid-stream. Whereas on 24 May 2014, the German conglomerate was supposed to put a firm offer on the table for Alstom, it instead simply asked for more information. Siemens was keen to learn more about the legal proceedings pending before the DOJ. It had an inkling that the penalty inflicted by the DOJ would be a financial blow for the company that it was hoping to acquire. Siemens knows only too well, having been previously under FCPA investigation in 2006 for corruption in the United States. It was accused of having paid bribes in Argentina, Venezuela, the United States, China, Vietnam and Iraq. A system of corruption on a scale comparable to that of Alstom.

To settle this case with the DOJ and the SEC in 2008, Siemens pleaded guilty and agreed to pay a ground-breaking fine of nearly $800 million and severed all ties with its CEO, Heinrich von Pierer, who agreed to pay €5 million to his former company to avoid being sued for managerial misconduct. The case did not stop there, as in 2011, the American judiciary indicted eight of Siemens' former senior executives and issued international arrest warrants against them. This scandal has been a millstone around the group's neck for over ten years now. The repercussions it has had on Germany have already cost it more than €1.5 billion. Against this backdrop, its reluctance to relapse into a similar scenario to Alstom is totally understandable.

Conversely, GE appeared to be quite safe from the DOJ. The US titan even offered to rescue Alstom from the brink of disaster. In the agreement it submits to the French group, there is a section stipulating that in the event of a takeover, the US entity will assume all its legal liabilities. In other words, GE undertakes to pay the fine that Alstom will have to pay to the DOJ. It amazes me that such a provision would have been negotiated. If a company is not allowed to take over the penalties inflicted on its employees, the most rudimentary logic holds that it is just as impossible for one company to substitute itself for another, especially in the case of a partial takeover. It is therefore surprising that the DOJ did not object to this clause as soon as it was made public in June 2014.

Nevertheless, GE's pledge to pay in lieu of Alstom was a compelling argument that Siemens could not match. Early June 2014, it was impossible for anyone to know the amount of the ultimate fine. Alstom didn't plead guilty until six months later, on 22 December 2014. Just who would agree to sign a blank cheque for an amount upwards of $1 billion? No officer in the world would ever obtain the approval of their board of directors and shareholders to make such an offer. That is certain. Yet again, neither the financial press nor our political elites appeared to perceive any incongruousness here. GE and Alstom's communications teams had pulled out all the stops. Yet the question is crucial: how could GE commit to an unknown amount that could represent up to 10 per cent of its bid price? Could it be that GE possessed insider knowledge that Siemens did not have? General Electric had been involved for months behind the scenes in Alstom's negotiations with the DOJ. As reported by the *Wall Street Journal* on 4 February 2015, 'GE reviewed the documents related to the settlement with the Justice Department "at all stages in their preparation and negotiation," Alstom's lawyer Robert Luskin of Squire Patton Boggs told a judge at a December plea hearing.'

Katty Choo, an ex-federal prosecutor specialized in the prosecution of financial crimes and now the head of GE's 'investigation and anti-corruption' unit, also later attended Alstom's plea hearing.

So at this stage it was all played out by current or former DOJ prosecutors.

Early in June 2014, four thousand miles from Paris, I watched the latest twists and turns of the Alstom sell-off with a feeling of total hopelessness. It left a bitter taste in my mouth. I feel like I've been ambushed, but not just me, the whole of France too. Would it not have been better to break my silence and ask my family to publicize my story so that the whole world knows the score? Perhaps. Without doubt. I did think about it for a while. Clara could have alerted investigating journalists or responded to requests from Montebourg's envoy. But would it have served any purpose? How do you tackle the DOJ, Havas, Publicis, GE, Alstom, Patrick Kron, François Hollande and Azéma as a group? It's known as fighting a losing battle.

The priority for me and my wife, my four children, my mother and father and my sister, is that I finally get out of this prison. So yes, maybe I am selfish by not speaking up, but I have been incarcerated for almost fourteen months now and, for the first time in days, I can see light at the end of the tunnel. I therefore prefer to keep my head down.

# 33

## The road to freedom

In Paris the case is closed, the decision has been taken. GE has successfully secured the deal with Alstom and all will be signed in a week's time. Meanwhile, Assistant US Attorney David E. Novick informs Stan that he can file a motion for my release. On 11 June 2014, I therefore begin my last day of detention, the 424th day. Tomorrow I'll be out.

My last day at Wyatt is the same as any other day. I get up at 6:50 a.m., have breakfast, then do one hour of gym with Alex with just a towel on the floor in a corner of the refectory and an hour's speed-walking in the confines of the small walk yard. This yard is like a microcosm of the prison; it is completely enclosed and covered.

I have been living under the constant glow of fluorescent lights for 250 days now. I happened to be unlucky enough to have been assigned to a pod with no daylight. They treat me the same as any other prisoner, no better, no worse. We are all mistreated in fact. Whatever his crime, a human being should not be prevented from breathing clean air and feeling the sunlight on his skin. He should not be treated worse than a beast. Such cruel treatment can turn a man insane and even vicious.

We were also banned from using the yard due to 'budgetary restrictions'. This 'prison capitalism', this race for profit at the expense of the most basic human rights, is degrading. It is not just about optimizing the profitability of prisons, increasing the number of inmates, making the maximum profit; there is another objective, i.e. exert enough pressure on the inmates that they crack and plead guilty as soon as possible, saving the DOJ the costs of a trial and, in the process, improving its already Stalinist statistics (98.5 per cent success rate).

This morning, a few hours before my release, I walk fast to evacuate my anger and my hatred of Wyatt and the US judicial system. I feel drained and exhausted. One of my fellow inmates joins me. Teka is Albanian, he has just returned to the pod after four days spent in Providence hospital. They removed a huge three-inch cyst from his throat, at his Adam's apple. This massive growth should have been removed much earlier, but Teka had to wait three long months before the administrative machine allowed him to be hospitalized. I witnessed him slowly wasting away. His ever-growing cyst blocked his oesophagus; he was unable to eat solid foods at the risk of choking. Since February, he had only been drinking soups, and since he was unable to breathe normally, he could no longer sleep at night.

He had to fill out dozens of administrative documents to be eligible for an outside consultation. He now has a huge scar on his neck. He looks like Frankenstein. He will not be able to turn his head for several days, but he is so grateful that the operation was a success. His doctor told him that he had removed a protrusion 'as big as a black pudding' from his throat. The surgeon was outraged at the delay in medical care he had experienced. I too am outraged. I have quickly learned that life does not have the same value inside as it does outside. As I start packing, I think back to some of my fellow inmates. India, a sixty-five-year-old of Indian origin, did not have this chance. He died a month ago because he received medical care too late. Then there was Kid, who couldn't cope with the psychological strain when the prosecutor made him an initial fifteen-year offer for a drugs affair: he hanged himself from his bed. He was only twenty-four years old. It was his first offence. Mark, whom I shared a cell with for seven months, remained in Wyatt for five years before being tried last December. He was planning to spend Christmas with his family. But a fortnight before his sentencing, the prosecutor said that the fraud was far more serious than initially estimated, and Mark was subsequently sentenced to twenty-five years in prison! Bob, married for forty years, lost his wife two months ago. The prison administration declined to escort him to the funeral in Boston, and instead proposed to him that the hearse be brought to the yard of Wyatt Prison so he could meditate for a moment, which he of course refused outright.

In view of all this, I should count my lucky stars that I'm not leaving Wyatt in a body bag. I just hope to get back to a 'normal' life soon.

Later, I will say farewell to my most faithful friends, Peter, Alex and Jack, with whom I have been spending all my days for one year now. We are among the very few 'white-collar offenders' in the prison. Just 10 out of 700 inmates.

Peter is starting to see light at the end of the tunnel. He has been incarcerated in Wyatt for more than three years. Outside, he acted as a 'transporter' by carrying cash for the Mafia in suitcases between New York and Las Vegas.

Jack, too, should soon be eligible for parole. He was on the front page of the US press. He was nicknamed 'little Madoff' by journalists. This sixty-two-year-old financier set up a 'Ponzi scheme', to scam investors in the United States. He skilfully pulled off a good deal with the DOJ. In total, he was 'only' sentenced to seven and a half years in jail.

The reverse is true of his co-defendant Alex, who was his subordinate. Alex refused to plead guilty. He tried to resist the system and insisted on going to trial. A fatal error because he risks being sentenced to a stiffer penalty than Jack. This is further proof for me that the US penal system is a game of Russian roulette. Alex is extremely worried as he awaits his trial. Throughout my detention, he was my closest friend. Before moving to the United States, he graduated from a business school in Marseille and speaks perfect French. He is a cheerful religious character in his fifties, who has managed to maintain his zest for life even behind bars. During the fourteen months I spent at Wyatt, Alex never failed to lift my spirits. He will remain my friend for life.

# 34

## Free at last!

Right until the last minute they gave me a hard time. When Wyatt's guards came to get me out of bed at 4:00 a.m. this morning, I was expecting to be released early. They then sent me in a cell van to the court in Hartford (Connecticut's capital). Once there, they locked me in a cell in the courthouse. Since then, nothing has happened. I've been languishing in this dungeon for almost eight hours, whereas all the formalities have been completed and there are no problems with my bail. Is there a last-minute problem? I have heard so many different tales in Wyatt that I'm no longer sure of anything.

And I'm not the only one who's languishing right now. My father, who is over seventy-five years old, has also been patiently waiting since this morning, sitting on a court bench with our friend Linda, who insisted on accompanying him. He has been hanging around in a lobby just a few yards from the jail where I am locked up. We're so close we could probably talk to each other.

Good sign, around 4:00 p.m. they bring me the clothes that they took from me upon my arrival at Wyatt. They're way too big for me now and I look ridiculous. Finally, the door opens, and at the end of the hall, my father and Linda are waiting for me; they stand up and welcome me with open arms. I taste freedom.

We hug each other like crazy. My father seems to be in good shape. In any case, he looks far more robust than he did a few months ago when he came to visit me at Wyatt despite my pleadings not to. At the time I thought how small and frail he looked.

His back was racked with pain, he was bent double, was breathless and had to use a walking stick. He was obviously in pain, yet he still braved the journey across the Atlantic to see his son for just two hours behind a glass pane. And he is also here to greet me now. Clara had to stay in Singapore to complete the formalities for the end of the children's school year before organizing the move to France. But she will soon be here with Léa, Peter, Gabriella and Raphaella. The whole family will be here in a month's time and we will have a few weeks' vacation together.

No matter how much I value these first hours of freedom, I know that my ordeal is far from over. I have only been released on bail for an initial period of two months. I have been ordered to remain in the United States. I will be staying with my friend Tom in Connecticut and I will only be allowed to travel to three other states: Massachusetts, New York and Florida, where I plan to take the children and Clara in a month's time. Since Tom has joint custody of his children, my father and I rent camp beds. We sleep in the living room of Tom's house.

My memories of these first hours of freedom are fuzzy. Rather, I remember snippets of sensations such as the warm comfort of a hot bath, my first in fourteen months. The smell of grass, of trees, the wind gusting. As soon as the kids wake up, I Skype them and Clara. I see my kids for the first time in fourteen months. They have changed so much, it's unbelievable! We don't speak for long as they have to go to school, but it's so wonderful to finally see and hear them.

I also remember sitting for hours in Tom's garden, gazing at the open sky, almost blinded by its expansiveness. I have noticed that after being locked up in small confined spaces, my field of vision has shrunk. It took me a few days to adapt and be able to look into the distance and clearly distinguish the horizon. In jail, my senses had withered due to smelling, seeing, touching, eating and hearing the same things for over a year.

I spent these first few days taking long walks in the forest. Sometimes, when he is feeling up to it, my father joins me. My parents separated when I was quite young. I have spent more time with my mother. In June 2014, I truly discover who my father really is. I ask him to tell me about his life, the company he founded, his commercial exploits in Russia, and I encourage him to record some short videos for his grandchildren. I spend the rest of the time online. I had naturally been deprived of this in jail. I collect a maximum number of documents and I read all the articles published on the sale of Alstom to General Electric. I compile, classify and list this data. I am determined to launch my own counter-investigation.

Mid-July 2014, Clara and the children finally join me.

I go to pick them up at JFK airport with a knot in my stomach at the thought of setting foot in that airport again. The children do not hide their joy or their surprise when they see me. I have lost almost twenty kilos. I would almost look younger if I hadn't still got that greyish look that three weeks of freedom have failed to erase. To be frank, I still have a 'dirty mug' – a true jailbird face.

I soon see that Pierre is now almost a head taller than Léa. My two youngest, Raphaella and Gabriella, are so excited that I mustn't let go of their hands, lest they start screaming or scolding me. I reluctantly have to leave them at night to comply with the conditions of my bail and return to Tom's house where I spend the night. My family is staying with our friend Linda. We are so happy to be going to Florida a few days later. I obtained permission from the judge to spend three weeks there. This is our first real taste of happiness. We are staying in a hotel apartment by the sea. Léa, who plans to become a swimming champion, swims three miles every morning with the Miami lifeguards while the two youngest babble away playing happily in the sand and Pierre hurries us onto an amphibious boat, a local attraction. These boats, after rolling on the sand, plunge at high speed into the sea. Beach, sun, waves, just an ordinary vacation, though for me it is simply extraordinary.

Three weeks later, Clara and the children have to return to France, and I start a new legal battle to obtain the right to join them. Then comes some good news: William Pomponi, after resisting for over a year, finally agreed to plead guilty on 18 July, one month after my release from Wyatt. This confirms that my ongoing detention had nothing to do with Pomponi's criminal situation and is entirely related to the deal between GE and Alstom. His counsel must have done a good job as, unlike me, he only had to plead guilty to one count.

Since I no longer have to wait for his trial, I am now hopeful that the DOJ will be more accommodating and allow me to return to France. Just as I was starting to relax a bit and get used to being with my family again, my hopes are quickly dashed. I learned from Stan that the prosecutors are now blocking me, due to Hoskins. They're trying to pull the same stunt on me as they did last time using Pomponi. I therefore cannot be sentenced until Hoskins pleads guilty or goes to trial. If they continue to play this game, indicting other people, it will never end. I risk remaining in detention for months or even years pending judgment, while our finances and those of our friends are blocked. How can I live like this and try to rebuild my life with all this hanging over my head? How do you persuade a future employer to hire you knowing that you may have to return to jail at any moment, for maybe up to nine years? It is impossible, and yet I have to start working again. I am only forty-six years old.

Moreover, Hoskins may well give the DOJ a run for their money. He only worked for three years at Alstom, which he left on 31 August 2004, just after the Tarahan contract was signed. And during his three years as International Network senior vice-president for Asia, he never set foot in the United States. Under these circumstances, his counsel were able to raise many legal issues. For instance, what competence does an American court have to try a British citizen for acts of corruption in Indonesia, who has been retired for many years and who only worked in France for three years, during which time he never set foot in the United States? Are the facts not time-barred? Not to mention other more technical points.

On the merits, I have no objection to his approach, I totally agree with it. I would have done precisely the same thing if I had been allowed to defend myself without being incarcerated. Nevertheless, I'm in a bind once again. As usual, I curse my lawyers, I get angry, and as usual, Stan presents the same Hobson's choice.

'If you don't follow our advice and insist on being sentenced now, Prosecutor Novick will have you sentenced to ten years.'

Always the same threat! I am furious to be cornered in this way, while Kron is at large out there, without me knowing what deal he has concocted with the American authorities.

Stan tries to convince me that even if an agreement exists, it has nothing to do with my case. Faced with this, I write an email, which my lawyers send on 18 August to the prosecutors. I need to check it out for myself.

The DOJ did not reply. Yet my request was perfectly legitimate. It falls within the scope of what the Americans call 'the discovery', a process that allows an indicted person to gather exculpatory evidence for his or her defence. I asked Stan about the Department of Justice's lack of reaction.

'It is possible that this agreement exists,' he tells me. 'But you'll never see it. The DOJ is under no obligation to give it to you, and if it is a confidential agreement, the DOJ is not authorized to supply it to you, or even to acknowledge that it exists.'

'Okay, but at least if it doesn't exist, the DOJ could have told me. They are allowed to, aren't they?'

'They haven't answered you and they can't lie in writing, so draw your own conclusions.'

It is the end of August. Clara and the children are returning to France. It's heart wrenching to see them go. I have no idea when I will see them again.

# 35

## Back to France

I've had enough, I refuse to remain stuck indefinitely in the United States without the right to work, without my family, totally cut off. I just can't do it. I don't give two hoots about Hoskins' legal battle. I just want to go home. I don't want to hear my lawyers any more.

I therefore urge them to obtain an extension of my release on bail as soon as possible. My intransigence finally pays off. The prosecuting attorneys start to loosen their grip. And this negotiation, like all the others, reveals the same bottom line: dollars. They eventually authorize me to return to France, on condition that I raise my bail amount. My father's friend, Michael, who has already generously agreed to put up his family home as collateral, adds another $200,000. I am also prohibited from travelling outside of Europe (unless specifically authorized by a judge) and once back in Paris, I must send an email to an American probation officer on a weekly basis. I naturally agree to all these conditions.

Before my departure scheduled for 16 September, however, I must fulfil a promise I made. The sentencing hearing of my former cell mate Alex is scheduled for early this month in Boston.

I guess there won't be many people in the courtroom to support him and he'll be happy to see me. In fact, only three of us attend, the Greek consul, a cousin of his who made the trip from Athens, and me. When Alex walks in, handcuffed, he gives me a wide grin. The hearing only lasts thirty minutes. The prosecutor is having a field day because Alex had the courage to defy him by going to trial. My dear friend, like all the accused, reads out a prepared text in which he asks the whole world for forgiveness. His case was heard, and the verdict was pronounced: one hundred and two months in prison, which equates to eight and a half years. He got one more year than Jack his alleged partner in crime, who was the ringleader. Alex is devastated. Including the fifty-four days a year of remission for good behaviour, he can expect to be freed in 2019. One last look, one last wave and they take him away.

I haven't left yet. And the US authorities do not fail to amaze me, right to the end. The DOJ has encountered a new problem. It cannot agree with Homeland Security on the type of visa I will need when I return to the United States for my trial. Why should I care?

'It's not that simple,' Stan explains. 'They want you to "take the risk".'

'What risk are you talking about?'

'The risk that they may not find a way to get you back into the United States legally. If they cannot come up with an administrative solution, they will seize the bond and the DOJ will consider you as a fugitive.'

'That's sheer madness! What can I do about it?'

'If you want to leave as planned on 16 September you will need to sign a document acknowledging your liability and disclaiming theirs, if they cannot find a solution.'

Back to the Kafkaesque nightmare. Before I decide, I need Linda and Michael's authorization, as they risk having their homes seized. When I tell them about the inextricable situation I am in, they can't believe it. I also ask the opinion of Jérôme Henry, the vice-consul of Boston, who is stunned. They all encourage me to go anyway. 'We'll involve the Embassy,' Henry promises me.

So that's it, I'm going home. I am very nervous about getting back on a plane and passing through the controls again. I'm afraid that I'll be arrested again at the last minute … It's only at take-off that I finally unwind.

On 17 September 2014, I land in Paris. It is now 493 days since my arrest. My father is there to welcome me. We head straight for the house. I arrive just in time to pick up my two little girls from school. I will never forget the look of amazement on their faces and their wide eyes when they spotted me in front of the school entrance.

As the days go by, I try to resume a normal life. Well, almost. It's not like it used to be. Things that were natural, simple or obvious before, have now become difficult. I have to re-programme myself, accustom myself to family habits and routine, regain my place as father and husband. It turns out to be a lot harder than expected. Prison leaves permanent scars. I am now an unemployed father and husband.

On 2 October 2014, I register at the job centre for the first time in my life.

# 36

## My meeting with Matthieu Aron

He left me a voicemail a few days ago. He said he wants to hear my story and is seeking information on Alstom's sale to General Electric. I'm very suspicious. Before agreeing to meet him, I check him out. My sister Juliette too. She tells me that she has often heard him on the radio, that she has already read his books and thinks he is 'reliable'. However, I am worried as I am out on bail, I have not yet been sentenced and I live with the permanent threat of being put back behind bars. If the United States prosecution finds out that I have been talking to the press, they could make me pay dearly. It's been three weeks now since I arrived back on French soil, I should feel happy and relieved, but I have trouble finding peace of mind. I'm permanently on my guard.

When I was locked up in Wyatt, I invented a secret code for communicating with close family members. We decided on the title of a book that contained about fifty characters to whom we assigned numbers and letters. Using this encryption, I believed I could exchange 'secure' letters. As it happened, the mechanism I had devised proved quasi-impossible to memorize. We ended up never using it.

What will I get out of meeting this journalist? Is it a set-up? Others have tried to contact me or Clara, and so far we have sent them all packing. In the end, even if I live to regret it, I finally accept a meeting for 9 October 2014.

I decided that we would meet in the market square of the old town of Versailles. That way, I can see if he has sneaked a photographer in with him. Well, he appears to be alone. All the same, I remain suspicious. As I approach him, I stay behind the wheel, ask him to quickly jump into the car, and without even taking the time to say hello, I drive off. I turn several times in the town centre to check that nobody is following us, before heading at full speed towards the Château de Versailles. I planned to stroll with him in the vast gardens of the Château so that I could easily spot anyone watching us.

With hindsight, I realize that I had met Matthieu Aron (who at the time was a journalist at France Inter) in rather preposterous circumstances. This unusual introduction didn't appear to offend him; on the contrary, it seemed to amuse him. We walked together in the Château gardens the whole afternoon.

Since I left Wyatt, I have got into the habit of walking at a brisk pace, every day for hours. It eases my anxiety and is synonymous with sport. We walk and I talk. I start off slowly, then it all comes pouring out. It's somewhat liberating. He did an incredible job of putting me at ease and I found myself relating the whole story: jail, the shackles, the humiliation, my despondency, my fears, my family's distress, the criminals, the screams, and Alstom.

In a piecemeal way, I described to him how some international markets were obtained using "consultants", the window dressing, i.e. anti-corruption procedures implemented by the company's management to conceal corrupt practices, and how Alstom had ultimately betrayed me. Most importantly, I told him that I was convinced the Americans had framed us and had manipulated Kron, who chose to sell the company rather than serve a heavy prison sentence.

I quickly realize that Matthieu Aron already knows part of my story and is fully aware of the forced sale of Alstom to GE. However, the press displayed considerable discretion on the subject.

Only one short article was written about me, headed 'Alstom's ill-fated executive. How Pierucci fell into their trap', written by the journalist of the *Journal du dimanche*, Bruna Basini in July 2014.

It was a story that wouldn't have captured anyone's imagination.

Matthieu Aron told me that he had been alerted by an Alstom insider. A senior executive from the company's inner management circle wanted to meet him and then revealed the behind-the-scenes legal aspects of the operation. For this senior manager, there is no doubt that Patrick Kron was pressured into selling to the Americans. Many other Alstom managers were equally convinced of this. He also had a number of interviews with certain political leaders. All of them expressed their indignation at the under-the-table hawking of this French industry icon. Why then did they not raise the alarm? For a very simple reason, he explained to me: none of these people would agree to being recorded or would even accept being quoted.

I tell him that unfortunately the same applies with me. Coming out into the open would certainly imperil my legal situation. I ask him for his absolute discretion about our meeting today, which must remain secret. He undertakes to do so and keeps his word. This is how our cooperation started. We ended up meeting several times, both of us searching separately for evidence and clues to shed light on the economic warfare waged by the United States against Alstom. The journey was long and we had to be patient, very patient.

Right now, early October 2014, we say our farewell after a very long walk. Matthieu will try to gather witness accounts and I will search to uncover secrets or gather documents to back up our joint research effort.

Three weeks after our meeting, I am about to leave for the United States. My authorization to stay in France, which was limited to eight weeks, is ending. At the last minute, nine days before my departure, I receive a message from Stan: 'You can stay in your country until 26 January 2015. The proceedings against Lawrence Hoskins have been delayed.' So I have three months of respite.

# 37

## Speak or remain silent

'We will never be able to offer you a post that matches the role, level of responsibility and compensation you had before,' replied the employee at the job centre. I hadn't even told her I had just got out of jail and risked going back inside for several years. So here I am, at forty-six years of age, unemployed, unable to apply for a job in an organization (as I am out on bail), and unable to hide my past (as any recruiter, client or partner could type my name on Google and discover my indictment). I can't even live on my savings or pension, as they all went into the bond and we have four children to support. Fortunately, in September Clara found a job in Paris shortly after her return from Singapore.

Rather than moping around, I decide to use my newfound experience. I spent months reading, re-reading, and analyzing in detail all the FCPA cases. During my probation period in the United States I created a database, collating thousands of documents. In parallel, I also studied the anti-corruption laws of France, England, Germany, Switzerland, Spain and Italy. However, after a quick overview, I realize that in France the compliance consulting market (corporate ethics) is almost exclusively dominated by American companies. Whether audit firms, large law firms, or economic intelligence agencies, almost all of them are Anglo-Saxon. This is not surprising, since the compliance market was born in the United States, which has made it a global business. Except that, in this case, we are dealing with national security and economic sovereignty. Just look at the list of companies being prosecuted by the DOJ, whether Alcatel in telecoms, Total and Technip in oil, or Alstom in energy. And this is just the tip of the iceberg.

In September 2014, an American site 'FCPA Blog' listed those French companies at risk: Airbus, Sanofi, Vivendi, Société Générale. Many blue-chip CAC 40 (the main French stock market index) companies were operating, sometimes unknowingly, under the threat of FBI investigations. And yet by 2014 not one of the major French law firms had an anti-corruption compliance department. Only two associations – the Cercle Ethique des affaires and the Cercle de la Compliance – tried to convey the right message to companies.

On a very small scale, I therefore decided to embark on the venture myself, by creating Ikarian, a small consulting business with two objectives: first, to raise awareness among top managers, and second, to offer them a range of services, including upgrading their compliance procedures, creating their corruption risk map and verifying the integrity of their stakeholders (distributors, intermediaries, suppliers, customers). For months I worked on creating my own tools to ensure I was the best performer on the market. At the end of 2014 when I first started out, I set myself a few ground rules: no public intervention and no talking openly to journalists, who are all over me like a rash since I returned to France. Nor must I publicize my new job (no website, no promotion). Obviously, it's harder to drum up clients in these conditions, but that was the price to pay.

Another of my objectives was to raise awareness among political leaders that French anti-corruption laws needed to change. I cannot just stand by and watch our companies being literally held to ransom by the United States Treasury. Other European enterprises caught on before we did. Following the BAE Systems case in 2010, the British adopted their own anti-corruption law, known as the UK Bribery Act. So why can't we follow suit in the wake of the Alstom case? I received some excellent assistance on this issue. This was provided first and foremost by Paul-Albert Iweins, former president of the Paris Bar Association and of the National Bar Council. Like my lawyer Markus Asshoff, he is also a partner of the law firm Taylor Wessing. Over a two-year period, Paul-Albert was one of the most active lobbyists for regulatory change in France, which culminated in the adoption of the Sapin II law in December 2016 reforming anti-corruption measures. After that, Didier Genin, one of my most loyal friends, introduced me to Éric Denécé, a former military intelligence officer who later became an economic intelligence officer.

Director and Founder of the French Centre for Intelligence Studies (CF2R), Éric Denécé, together with journalist Leslie Varenne, published a damning seventy-page report in December 2014. It was entitled *The Alstom Affair: American Racketeering and State Resignation*. The two authors denounce Patrick Kron's fallacious arguments, the powerlessness of the French state, and above all express their alarm at the threat to France's nuclear independence: 'In terms of turbines for surface vessels and for nuclear submarines for the French Navy,' writes Denécé, 'General Electric, after having absorbed Alstom's energy sector, now enjoys a near monopoly position rendering our naval fleet hugely dependent on its deliveries. And, in terms of space surveillance, we are also divesting the Alstom Satellite Tracking Systems subsidiary, which supplies our armies, notably the Directorate of Military Intelligence (DRM), and contributes to the effectiveness of our nuclear deterrence through the constant surveillance of Allied and enemy satellites.'\*

At the end of 2014, Éric Denécé, who has been advising me on my case since I returned to France, is surprised that I have not been contacted by the Economic Intelligence Unit of the French

---

\* Extract of an interview with Eric Denécé from *L'Humanité* dated 19 July 2014.

Ministry of Economy. With my consent, he informs the officials in charge. His counterpart believed I was still in the United States ... Odd for an intelligence expert not to know the score. I was quickly summoned to the Ministry of Economy for a full debriefing and was received by the head of the Economic Intelligence Coordination Unit, accompanied by the research laboratory director Claude Rochet and a legal expert. They tell me that they have clearly identified the DOJ's machinations to destabilize Alstom, but they lack crucial elements, which I am able to provide. I meet them three times over the next few weeks. Even if I don't feel that the warnings they tell me to give the minister in charge are of much use, at least I feel my cause is being heard and it is a great comfort to know that high-ranking officials understand the foul play I have been subjected to, which reassures me that I'm not insane or a conspiracy seeker. Meanwhile, I meet other people who are eager to have the truth exposed, including Marie-Jeanne Pasquette, editor-in-chief of the minority.com news site, which defends the interests of small shareholders in major enterprises. She also conducts a very detailed investigation on Alstom.

From that point on, I observed that many intelligence experts or economic analysts had clearly understood what was behind this sale. The political elite however adopted a 'laissez faire' attitude throughout and let it happen. On 4 November 2014, Alstom's Board of Directors unanimously authorized the signing of the Master Agreement with General Electric.

The next day, the recently appointed Minister of the Economy, Emmanuel Macron, who had just replaced Arnaud Montebourg (who left slamming the door behind him), endorsed the operation. Macron waived the state's right of veto, which would have allowed him to block a foreign investment of this scale in France. His predecessor had fought long and hard to get such a veto enacted. A few weeks later, he did however oppose the sale of the French online video company Dailymotion to the Hong Kong-based

PWCC led by tycoon Li Ka-shing, favouring a 'European solution'. We obviously don't agree as to what constitutes a strategic enterprise.

The last hurdle to overcome for completion of the sale of Alstom's energy division was approval by Alstom's shareholders. An extraordinary shareholders' meeting was held for this purpose on 19 December 2014. That morning, France Inter broadcast the first major enquiry into the 'backdrop of the Alstom sale', on prime time just a few hours before the shareholders' meeting due to be held in the salons of the Hôtel Méridien Étoile, Porte Maillot, in Paris.

Matthieu Aron finally convinced one of the company's senior executives to speak out while remaining anonymous. The statement was ravaging.

'Everyone at Alstom's headquarters is well aware that the legal proceedings instituted in the United States against Alstom were instrumental in the decision to sell the energy business,' says the manager. 'The legal proceedings are the key to everything. It's an open secret.'

France Inter also draws on the statements of the vice-president of the National Assembly's Economic Affairs Committee, Daniel Fasquelle, one of the very few parliamentarians aware of the hidden implications of this affair.

'The Alstom case is an unbelievable hoax,' said the MP. 'We have deceived the French people. In no way have we saved Alstom and we should question ourselves about Alstom's problems in the United States and the ongoing lawsuits. The takeover by General Electric was a convenient way for Alstom to extricate itself from the legal trap set for it by the US judiciary.'

I remember listening to this radio broadcast on 19 December 2014. It was 7 a.m. and I was on my way to Porte Maillot to attend the meeting.

# 38

# Acrimony at the shareholders' meeting

I would have given my right arm to not miss this meeting, especially since they tried their utmost to dissuade me from attending. I did, though, deliberately try to keep a low profile by registering at the very last minute, just a few hours before the registration deadline. To no avail, however, as they spotted my name immediately.

The previous evening at precisely 8:50 p.m., I received an email from my lawyer Liz Latif.

'Hi Fred, we have been informed that Alstom shareholders are meeting tomorrow regarding GE. If you decide to attend, we advise you not to speak publicly and to remember that anything you say may be used against you by the DOJ.'

I am appalled at this message. Who on earth tipped off my lawyers? I reply to Liz, expressing my surprise.

'Thank you for your advice. But who asked you to warn me? The prosecuting attorney?'

She immediately replies: 'No, we are informing you voluntarily.'

This seems doubtful to me. How could they have known I had registered if no one had told them? I haven't heard from them in weeks. Am I to believe that, all of a sudden, they independently chose to reach out and caution me?

'Liz, I'm very surprised. I don't suppose for one minute that this shareholders' meeting was announced in the US press, so could you tell me who informed you of it? Was it Patton Boggs, Alstom's counsel? Or was it the DOJ?'

I end my mail with a touch of irony: 'Don't worry, I won't do anything that could endanger GE's takeover of Alstom. Best, Fred.'

Liz Latif does not respond. Our email exchange ends. It's 2:48 a.m. The shareholders' meeting starts in less than eight hours.

I take great pains to arrive early. The warnings given by my lawyers have not deterred me from attending, quite the contrary. I had no intention of speaking up anyway. I'm not stupid, I fully appreciate the consequences thereof. All I want is to stare both Patrick Kron and Keith Carr straight in the eye, defiantly. Now that I no longer need to exercise discretion, I decide to sit in the second row behind top management, right in the CEO's line of vision.

Seated next to me, I recognize the representatives of the majority shareholders, Bouygues and Amundi. Advisers of large investment funds are also present, but above all there are many private individuals who hold just a few shares, regulars at this type of meeting, most of them senior citizens. The general meeting was barely underway when my mobile phone started to vibrate. This time it is Stan sending me a message: 'Do not do anything that could compromise your situation!' It was precisely 10:32 a.m. in Paris, i.e. 4:32 a.m. in New York. Stan is on the alert in the middle of the night. He must be under some pretty intense pressure. I don't reply for the time being and concentrate on Patrick Kron's opening speech.

In the gallery, dressed in a midnight-blue suit and a mother-of-pearl and lilac herringbone tie, he is seated in a comfortable white leather armchair next to Kareen Ceintre, the secretary of the Board, and Keith Carr, group general counsel and member of the Executive Board. Keith Carr immediately spots me and does not take his eyes off me for the duration of the meeting, fearing I might decide to take the floor. I suspect that he is behind the message that Stan just sent me.

The CEO must now explain the details of the proposed sale to General Electric. Admittedly, the Board of Directors has already approved it. Thanks to the majority shareholders, the vote is a foregone conclusion. Moreover, many already expressed their views online prior to hearing Kron's statement.

This time, the big boss will not be able get away with approximations and declarations of intent, as he did during his press interviews. The reference document handed out to participants is highly specific. Upon reading it, many of the small shareholders or employee representatives must have been knocked for six. Contrary to everything that was reported hitherto in the media by both Alstom and General Electric, the French have indeed handed over the keys of the company to the Americans. The so-called 'alliance' widely touted by our political leaders was sheer illusion. It was merely a neat trick to deceive the public. The agreement that was signed was not a 50/50 partnership. In fact, in the first two joint ventures (Networks and Hydraulics) Alstom holds 50 per cent minus one share. In other words, GE is clearly in charge and even appoints the CFO.

In the case of the third joint venture, responsible for the nuclear sector, the arrangement is more complex. In view of the strategic interests at stake, the French state wanted power of veto. This is of no consequence however, as GE has a majority in terms of shares (80 per cent) plus voting rights. Ultimately, in all three joint ventures, the US conglomerate enjoys full powers in terms of organization, strategy and finance. In addition, between September 2018 and September 2019, Alstom intends reselling its shares in the three companies at a guaranteed price.* It all unfolds as if Alstom's withdrawal from the energy sector was staged in advance. So much for the 'alliance' so highly praised by the French government.

---

* Alstom sold its holdings in the three companies in October 2018.

Patrick Kron also announces that Alstom is in the process of finalizing an agreement with the DOJ. The company finally decided to plead guilty and pay a fine. The DOJ set the amount at approximately €700 million. Yet once again the final agreement does not match the projections. The Department of Justice refuses to let GE pay this colossal amount. This fine will therefore be paid by Alstom (or what remains of Alstom). I'm not surprised as I always considered this part of the agreement to be illegal. It's just rather bizarre that the US DOJ did not express its disapproval earlier on in June. I quickly conclude that they were in fact silent accomplices in June in the ploy set up by GE and Alstom, whose purpose was simply to eliminate Siemens.

But that's not all. Since Alstom will now have to pay €700 million more, it appears logical that the purchase price of €12.35 billion negotiated in June should be raised by the same amount to avoid penalizing Alstom. Contrary to expectations, though, Patrick Kron says that the sale price will remain the same. And to justify his point, he uses an argument that even a five-year-old wouldn't swallow: GE will buy back some additional Alstom assets for around €300 million. This then leaves a €400 million difference remaining which the CEO considers as negligible. In his words: 'They correspond to the standard adjustment margin of a deal of such magnitude.' This comment meets with strong reaction from some of the attendees.

'You are confusing the two', says Marie-Jeanne Pasquette of minoritaires.com, denouncing 'a fully fledged shell game'.

In a matter of minutes, the small shareholders realize that Alstom has actually lost nearly €1.4 billion, i.e. the €700 million fine, plus the €700 million that GE will no longer be paying. The icing on the cake is that the Board of Directors has proposed that Kron be granted an exceptional bonus of €4 million for bringing these negotiations to a successful conclusion. The Americans must be rolling about in stitches.

I have trouble containing myself. I want to get up and vent my anger. A €4 million bonus for selling off an industrial gem! Four million for allowing a system of corruption to thrive for more than ten years. Where else but France would such an aberration occur? In France, board members form such a tight clan that no one dares to speak out. This is in direct contrast to Germany, where in 2008 the iconic chief of Siemens was fired in complete disgrace, then sued by his own company, which had to pay a fine of $800 million to the DOJ. In France, however, Alstom's incestuous Board decides that Kron merits a bonus and this does not seem to bother anyone. No reaction from the financial press, no reaction from the Finance Ministry or the government, no reaction from the main investors, no reaction from Bouygues, Alstom's main shareholder, no reaction from the AMF (French Markets Authority) either. Only a few small shareholders speak up.

The first on the attack is a certain Bulidon, who has been haunting the shareholders' meetings for more than ten years. He cuts to the chase.

'Let's be clear on this. I'm going to be belligerent and vote against this resolution as it represents the sale of two-thirds of the business.'

Bulidon pursues his rhetoric in an increasingly bitter tone.

'You always promised you would never sell Alstom off in bits, yet that's exactly what you're asking us to agree to today.'

Then he goes in for the kill.

'And as a reward for such a "fantastic deal", the Board of Directors is awarding you a bonus of four million euros. Mr Kron, if you have any professional conscience at all, forgo this bonus and resign!'

Instead of answering, Patrick Kron half smiles. Having been head of the group for over ten years, he has seen and heard a lot. Nevertheless, he must admit that Mr Bulidon, despite being very upset, is not entirely wrong.

'The three joint ventures are what they are,' he admits. 'They are explained in the presentation document. However, GE will indeed be in charge of operational control, but this is essential and quite normal.'

The CEO therefore clearly acknowledges that he is selling, and that it is 'normal' for the buyer to control the property he has acquired. Nothing could be more natural. Then why under these circumstances has he spun the truth until this meeting today?

A market investor, René Pernollet, grabs the microphone.

'I heard this morning on the radio [in the enquiry launched by Matthieu Aron] that the US court proceedings were instrumental in your decision to sell. Is that correct, Mr Chairman?'

'Okay, okay,' mumbles Patrick Kron, 'it's easy for you to lecture me and yell, but if we made this deal with GE it was to avert the imminent danger of turmoil within the company.' Then, once again in a deceptively caring tone, he continues: 'Don't scratch your brains out looking for arguments that don't exist. It's a good deal, full stop!'

This investor was merely asking a legitimate question, whereas Kron attempts to convince us that it is a good idea to shoot ourselves in the foot by divesting ourselves of our thermal sector, which is by far the most profitable sector. A few seats from me, I notice Claude Mandard, representative of the investment fund reserved for Alstom employees. Calmly and methodically, he in turn delivers his own indictment.

'It's a total botch-up and immense waste of our industry. In addition, you stabbed us in the back. If the press hadn't leaked it, we would have been faced with a *fait accompli*.'

This time Kron can't contain himself and explodes: 'First of all, the press leak has got us into a lot of trouble! Yes, it's put us in deep shit! For God's sake stop thinking that we were planning the hold-up of the century behind everyone's back.'

His violent outburst does not quell the anger of the small share-holders. The questions continue to be fired at him, each one more incriminating than the one before. The CEO then takes out his calculator.

'The sale of Alstom,' he explains, 'will net us 12.35 billion euros, from which we have to subtract cash, our investment in the three joint ventures, the buy-back of shares, and payment of the impending fine to the US authorities.'

He doesn't have time to finish.

'Please stop your subtractions,' a new speaker interrupts him sharply. 'Just tell us exactly what is left in Alstom's coffers.'

I had wondered myself. And the result is simply astounding, since it amounts to a big fat zero! Not only have we sold a world-renowned industrial asset, but we have nothing to show for it. To get the full picture, each element should be examined in turn. The sale price amounts to €12.35 billion. But from this €12.35 billion, you have to subtract cash (€1.9 billion), investment in the three joint ventures (€2.4 billion), capital gains allocated to shareholders (€3.2 billion), not to mention the €3 billion of debt to be repaid, then post in the accounts the purchase of GE's rail-signalling branch (€0.7 billion), and last but not least, payment of the fine to the DOJ: €0.7 billion. Of course, to all intents and purposes the company is debt-free, but its balance is close to zero.*

This sale can only be described as downright lousy and is without doubt one of the most nonsensical industrial initiatives to have been taken in recent years. It's abysmal. Patrick Kron denies this is the case, dismissing the facts as 'absurd conspiracy theories'. (He has never deviated from this line, writing in *Le Figaro* on 24 June 2019 that: 'There is no relationship between the conduct of an investigation by the US Justice (DOJ) on a limited number of old corruption facts and the merger with General Electric . . . I never had a discussion with the DOJ about a possible project with General

---

* It was Patrick Kron himself who prepared this calculation during the shareholders' meeting held in June 2015.

Electric until the leak in the media made this public and led the DOJ to question the parties about the contents of the project and its possible consequences on the current procedure.')

Despite Patrick Kron's denial, dismissing the facts as 'absurd conspiracy theories', I remain convinced that the US proceedings must be at the root of Alstom's dismemberment.

During the shareholders' meeting, questions on this subject are fired at him continuously. He is even asked about my detention in the United States. He is very wary and prefers to hide behind his confidentiality obligation rather than answer the question.

'The American case is ongoing, so it is impossible for me to make any comments whatsoever; it is simply impossible.'

At this point, Jean-Martin Folz, one of the members of the Board of Directors, decides it is time to intervene. He is a close friend of Patrick Kron, they met when they were both at Péchiney. Moreover, Jean-Martin Folz has been chairman of Alstom's Ethics Committee since 2011. He looks outraged, and in a serious tone he claims that the accusations are unfounded.

'The operations revealed by the Department of Justice in the United States are old, or indeed ancient, and do not involve the current Board of Directors. Since becoming head of Alstom, Patrick Kron has done his utmost to ensure that the company conducts its business affairs in an appropriate and acceptable manner. He has spent the last ten years doing everything possible to meet this objective.'

Three days later, the chairman of the Ethics Committee receives the most scathing counterattack in response.

# 39

## The DOJ prosecution show

I wouldn't have missed this for the world. I didn't see it live but the images were gripping. On 22 December 2014, seventy-two hours after Alstom's shareholders' meeting, the US judicial hierarchy calls a high-profile press conference. Dozens of journalists were present, and American TV stations were also at the ready. The footage was then broadcast worldwide on social media platforms.

A camera follows the entry into the conference room of Deputy Attorney General James Cole, who is accompanied by Assistant Attorney General Leslie R. Caldwell of the Justice Department's Criminal Division. The latter is a battle-hardened expert in charge of a 600-strong prosecution force.

The two prosecutors, with their serious faces and persuasive air, well aware that they are partaking in a historic moment, take their places in front of a lectern. A huge American flag, symbol of US supremacy, dominates the background. James Cole is the first to address the journalists.

'We are here to announce a historic law enforcement action that marks the end of a decade-long transnational bribery scheme: a scheme that was both concocted and concealed by Alstom, a multi-national French company and its subsidiaries in Switzerland, Connecticut and New Jersey.'

And the prosecutor continues, slightly short of breath but impressed it seems by the importance of what he is about to announce.

'Today, those companies admit that, from at least 2000 to 2011, they bribed government officials and falsified accounting records in connection with lucrative power and transportation projects for state-owned entities across the globe. They [Alstom and its subsidiaries] used bribes to secure contracts in Indonesia, Egypt, Saudi Arabia, and the Bahamas. Altogether, Alstom paid tens of millions of dollars in bribes to win $4 billion in projects – and to secure approximately $300 million in profits for themselves.'

James Cole then takes the moral high ground.

'Such rampant and flagrant wrongdoing demands an appropriately strong law enforcement response. Today I can announce that the Justice Department has filed a two-count criminal indictment against Alstom ... charging Alstom with violating the Foreign Corrupt Practices Act by falsifying its books and records and failing to implement adequate internal controls.'*

And, finally, the prosecutor announces the long-awaited decision.

'Alstom has agreed to plead guilty to these charges, to admit its criminal conduct and to pay a criminal penalty of more than seven hundred and seventy-two million dollars, the largest foreign bribery penalty in the history of the United States Department of Justice.'

In a matter of seconds, the American prosecutor has just destroyed the entire defence of Patrick Kron and his board of directors. The American investigators were not digging up stuff from long ago. Nor did they focus exclusively on 'old or indeed ancient' facts as maintained by Jean-Martin Folz, Chairman of the group's Ethics Committee three days ago at the shareholders' meeting. On the contrary, the DOJ investigated the last decade (2000–11) and Patrick Kron was chairman of Alstom and responsible for its destiny since early 2003. He should therefore be held fully accountable. In most cases, it was the Swiss subsidiary, Alstom Prom, that was

---

* The DOJ only opened a formal investigation on the day Alstom pleaded guilty. As such, the incriminations exactly match the facts for which Alstom pleads guilty. The DOJ can therefore boast a 100 per cent success rate.

entrusted with the task of paying the 'consultants'. The FBI was able to obtain all the bank transfers that were listed in detail in the indictment. Faced with this indisputable evidence, Alstom Prom and its parent company Alstom SA had no other alternative than to plead guilty and agree to pay the colossal fine inflicted on them. Two of the group's other entities, Grid and Power, fared a little better. They managed to negotiate a 'Deferred Prosecution Agreement'. They have undertaken to clean up their ranks and implement an appropriate anti-corruption programme. At the end of this time period, if the DOJ considers that the target has been achieved, they will be exempted from criminal sanctions definitively.

At this press conference, James Cole refers several times to a cardinal point of the file. Contrary to what Patrick Kron consistently maintained, 'Alstom failed to implement adequate internal controls'. It was pure window dressing. On the face of it, Alstom's policy was flawless, but behind the scenes the illicit behaviour continued in total obscurity. Its compliance programme was nothing but smoke and mirrors. And this system, according to James Cole, did not result from mere negligence or individual misconduct; it was carefully planned, codified and developed to perfection.

It was described by him as 'breathtaking in its breadth, its brazenness, and its worldwide consequences'. Alstom's corruption scheme was sustained over more than a decade and across several continents.

Finally, the Deputy Attorney General, in his closing remarks, issued a much more comprehensive warning.

'Let me be very clear: corruption has no place in the global marketplace. And it is both my expectation – and my intention – that the comprehensive resolution we are announcing today will send an unmistakable message to other companies around the world. The Department of Justice will be relentless in rooting out and punishing corruption to the fullest extent of the law, no matter how sweeping its scale or how daunting its prosecution.'

James Cole then acknowledged the assistance of the authorities in Switzerland, Saudi Arabia, Italy, Indonesia, United Kingdom, Cyprus and Taiwan for aiding the FBI in this matter. He didn't forget anyone. Though he did not mention France. And yet in Paris, as early as 7 November 2007, an investigation (revealed by the *Wall Street Journal*) was launched for 'active and passive corruption of foreign public officials', but the judges, for whatever reason, didn't appear overly interested in it. Then, in 2013, the public prosecution service also opened a second judicial enquiry into corruption involving Alstom in Hungary, Poland and Tunisia. Once again, the case got stalled.

It was a different case in Washington, however, where such matters 'excite' the prosecutors. They immediately seized the opportunity to capture a new French multinational, after Total, Alcatel and Technip.

In total, the fines imposed on these four major French companies have netted $1.6 billion for the US Treasury. If we add to this the extraordinary penalty of $8.9 billion imposed on BNP Paribas in 2014 for breach of an embargo, the $787 million bill that Crédit Agricole had to pay in 2015, or the $1 billion paid in 2018 by Société Générale, the total amounts to over $12 billion. This is higher than the French judiciary's annual budget. Imagine for a moment what the government could do with $12 billion. Let us take an example: the 'Great Plan to Eradicate Poverty' presented by Emmanuel Macron in September 2018. It amounts to €8 billion.

Back to Washington. Each prosecutor takes his turn and gives a dazzling performance. Next, the Department of Justice's unstoppable Leslie Caldwell takes the floor, to explain the investigations conducted by her crew. A press release was issued by the Department of Justice to accompany the press conference. It filled in some of the details on the breadth and depth of the investigation. Other details I later discovered myself, having been put on the scent by Caldwell's remarks.

It transpires that, in Saudi Arabia, as part of the Shoaiba project, an oil-fired power plant built on the Red Sea,* the company paid forty-nine million dollars in kickbacks (bribes) and used a sophisticated network of external consultants, all of whom were given code names such as 'Paris', 'Geneva', 'London', 'quiet man' or 'old friend'. Their mission was to bribe officials at the state-owned and state-controlled Saudi Electricity Company. The group also didn't hesitate to bribe a foundation for Islamic education assistance. In Egypt, from 2003 to 2011, Alstom engaged in corruption to obtain contracts from the Egyptian Electricity Holding Company (EEHC) by bribing Managing Director Asem Elgawhary of the joint venture company that EEHC had created with the American Bechtel.† Also, in the Bahamas, in an attempt to sell equipment, a consultant hired by Alstom bribed a member of the Bahamas Electricity Board of Directors. Finally, between 2001 and 2008, Alstom admitted to having paid bribes to Taiwanese officials to obtain contracts for the Taipei underground rail system.

At this press conference, Leslie Caldwell sought to justify the record-breaking fine imposed on the company.

'Through Alstom's parent level guilty plea and record-breaking criminal penalty Alstom is paying a historic price – as it should – for its criminal conduct and for its efforts to insulate culpable corporate employess and other corporate entitities. Alstom did not voluntarily disclose the conduct and did not agree to cooperate with the government – and in fact deliberately did not cooperate with the government – for some time.'

---

* The second phase of which was launched in 2004, well after Patrick Kron took over as head of Alstom.
† Which was not in any way prosecuted.

Finally, she utters a phrase that I have been recalling ever since.

'Indeed, it was only after the department publicly charged several individual executives that Alstom finally, and reluctantly began cooperating.'

A confession at last! The Head of the DOJ has publicly admitted that I was used as leverage on my company. I wasn't crazy or paranoid. I was used as a chintzy scarecrow to scare Alstom's management into cooperating with the FBI. There is a lot I can say about this justice system that, to successfully pull off spectacular deals, uses individuals as pawns. However, what I find even more shameful is the attitude adopted by my company. According to Leslie Caldwell, Alstom's management had refused to cooperate, contrary to its claims. In light of this, why didn't they warn me in April 2013 of the risks I was incurring by travelling to the United States? Why did Keith Carr throw me into the lion's den so to speak? I remember perfectly well his remarks made only a few days before my departure: 'You have nothing to worry about, everything is under control.'

How should I construe what he said that evening in Singapore? Did he knowingly sacrifice me by 'handing me over' to the DOJ? Or was it just crass incompetence on his part, with him believing he had fooled the Americans? I have asked myself this question repeatedly.

I favour this second theory. I think he was more incompetent than malicious. Though I could be wrong. Listening to Leslie Caldwell, I also wonder what would have happened if Alstom had adopted a different strategy vis-à-vis the DOJ? What would the legal outcome have been if management had admitted its misconduct in 2010? I know you can't rewrite history but three sensible theories spring to mind: (1) the fine would undoubtedly have been lower; (2) the group would probably not have been dismantled to this extent; and (3) the DOJ would not have had to arrest me. It should be noted that the DOJ did not make any arrests in 70 per cent of FCPA cases like, for instance, the ones of Marubeni, Total, Technip, BAE, and in many other cases similar to that of Alstom.

The scenario could have been entirely different.

And at that time, I didn't know what Matthieu Aron was going to discover during his journalistic investigations. He was able to gain the trust of Fred Einbinder, Alstom's former general counsel. The American lawyer who has been based in France for some thirty years was first head of Vinci's legal department before holding the same position at Alstom, where he remained until the very end of 2010 before being cast aside to make way for Keith Carr.

According to Fred Einbinder, Alstom's real legal troubles began in the mid-2000s.

It was in Switzerland that the corruption machine first started to spin out of control. In 2004, KPMG, the financial auditors working for the Swiss Banking Commission, audited Tempus Privatbank AG, a small private bank. The chief, Oscar Holenweger, was remanded in custody on suspicion of laundering money for the South American drug cartels. During a search of his secretary's home, investigators discovered that Oskar Holenweger also transited funds for Alstom in Lichtenstein, Singapore, Bahrain and Thailand. The transactions were copied out by hand to avoid leaving any electronic traces. The Swiss were to investigate for several years, before focusing on the group's Swiss subsidiary. They also transmitted information to their allies such as France, the UK and the United States. Though the investigation initiated in 2007 in Paris remained at a standstill, it was quite different elsewhere. First, in Switzerland, fifty officers of the Office of the Attorney General of the Confederation and the Federal Judicial Police carried out a series of raids in Baden, the Zurich region and central Switzerland. The Swiss even launched an appeal for witnesses by setting up a hotline to obtain testimonies against Alstom. In England, the British anti-corruption unit known as the Serious Fraud Office (SFO) struck a major blow on 24 March 2010. The code name of this operation on the other side of the Channel was Ruthenium, a metal of the platinum family, deemed very resistant, but which can be brittle at room temperature. The British police deployed huge resources to 'break' Alstom.

They mobilized 150 inspectors and searched the homes of the three managers of the group's British subsidiary: its managing director, its financial director, and its general counsel (the latter died from a heart attack the day after he was placed in custody). At the same time, in the United States, the DOJ prosecutors also got hold of the file.

At the end of the late 2000s, the American, Swiss and UK authorities decided to coordinate their investigations. The Swiss investigated contracts concluded in Latvia, Tunisia and Malaysia. The British focused on India, Poland and Lithuania. The Americans, on the other hand, were able to conduct investigations wherever they wished under their extraterritorial law, so they reserved the 'rest of the world' for themselves. Not to mention the Italians who, like the Brazilians, also triggered judicial investigations against Alstom, and finally, the World Bank, which suspected the group of corruption in connection with a contract in Zambia. Fred Einbinder, the then general counsel of Alstom, remembers the atmosphere that prevailed during this period at the group's Paris headquarters.

'It felt like we were being hemmed in wherever we turned.'

And this legal practitioner made one discovery after another.

'In light of my role, I was able to access the Swiss proceedings, and I studied all the contracts. I read and reread them for six to eight hours a day. I must have examined between one hundred to one hundred and fifty contracts. I would estimate that all the markets were obtained using bribery, whether on a small, medium or large scale.'

To deal with this legal tsunami, the general counsel set up a task force of legal advisers and lawyers. There were so many people involved that, in order to navigate his way around, coordinate and ensure the team worked effectively together, he had to draw up an organizational chart. Dated 26 November 2010, it included the names of English, Swiss, Brazilian, American, French, Polish and Italian lawyers. A total of thirty-nine of them were working on Alstom's corruption cases, but most senior managers like myself were unaware of this at the time.

'On the French side,' remembers Fred Einbinder, 'it was quite tricky. The company had several lawyers, but one of them, Olivier Metzner (a hotshot lawyer, now deceased), soon gained the upper hand. Working meetings henceforth took place at his office. This bothered me because Metzner was also Patrick Kron's lawyer, meaning there was a blatant risk of conflict of interest.'

But Fred Einbinder had other concerns. He was obsessed with the American investigation, which was slowly turning into a nightmare. Having spent the early part of his career in compliance firms, he was very familiar with the brute force employed by the DOJ.

'They made contact with us during the first quarter of 2010. The message was simple: you are a *target*. In short, they were letting us know that we were under investigation, and they were asking us to cooperate.

'This is how the DOJ operates. They approach large companies and offer them the chance to cooperate,' says Einbinder.

'Either you cooperate thoroughly, waive the prescription period, conduct an internal investigation yourself, and agree to self-incriminate, whistle-blowing on your own employees; or you refuse to make a deal, but you then have the FBI hard on your heels.'

This judicial system is the antithesis of our legal precepts here in France. In France, no lawyer would advise his or her client to hand over incriminating evidence to the prosecution. On the contrary, clients are advised to deny or conceal any incriminating evidence.

But Fred Einbinder had an Anglo-Saxon background and understood the need for coming clean, which is why, from the beginning of 2010, he repeatedly advised Patrick Kron to accept the rules set by the DOJ.

'At first Alstom's CEO ignored the advice, which merely angered him. He didn't want to be told; he was angry. He refused to admit our culpability. He even wanted to sue the prosecutors; it was getting out of control. I kept pushing and pushing and telling him to come with me to Washington.'

Eventually, in April 2010, Patrick Kron travelled to the United States with Fred Einbinder to meet with the law firm Winston Strawn, a specialist in corruption affairs. The introduction went well and Patrick Kron agreed to entrust the case to their lawyers at their headquarters in Chicago. They proceeded in the customary manner by conducting an insider investigation at Alstom. At that time, Patrick Kron thought he had merely authorized an audit, the sort that is regularly carried out in large groups. When he found out a few months later that American lawyers had conducted extensive investigations, he didn't hide his annoyance. They notably interviewed many managers, prompting them to reveal any corrupt or fraudulent practices.

On 10 December 2010, Winston Strawn sent a letter to Fred Einbinder and Patrick Kron recommending that they cooperate with the DOJ as soon as possible. In the course of their investigations, their legal experts had already uncovered the payment of bribes in Saudi Arabia, and they were confident that the FBI would do the same. At that point, the CEO made a radical decision and fired this law firm that appeared way too inquisitive and aggressive for his liking. He thanked his general counsel Einbinder by offering him a simple advisory role for a year to ease him into his retirement and appointed Keith Carr in his place.

How did Alstom subsequently manage this affair in-house? It's a mystery. Keith Carr* proved to be very discreet and hardly any information was leaked. The company probably hoped to be able to avert the storm. It is true to date that the European prosecutions have had little impact. In England, the sentences regarding the bribing scheme in Lithuania came out in December 2018: three of my ex-colleagues got sentences ranging from thirty-one months to fifty-four months of prison. Alstom Power Ltd was ordered to pay £18 million. In a parallel case, on 10 April 2018, Alstom Network UK Ltd were found guilty of 'one count of conspiracy to corrupt' for making corrupt payments to win a tram and infrastructure contract in Tunisia. In Switzerland in 2011 the company admittedly received a penalty order and a fine of 2.5 million Swiss francs plus a compensation claim of 36.4 million Swiss francs, but this remains a relatively small penalty. A year later, in 2012, the World Bank blacklisted two Alstom subsidiaries (including Alstom Switzerland) for a period of three years and ordered the group to pay a fine of $9.5 million.

As for the Norwegian Sovereign Fund (the world's largest investment fund), it preferred to withdraw from Alstom's shareholding in 2011, due to 'endemic corruption problems'. These sanctions undoubtedly tarnished the company's reputation, but at no time did they threaten its survival. Did Patrick Kron really think he could get away with it in the United States? I paid the price for his fatal error of judgment. Not to mention the Alstom employees and the people of France, who have witnessed one of their rare and strategic industrial giants vanish from the scene.

* The authors contacted Keith Carr, who refused to comment.

# 40

## Alstom's guilty plea

The American judicial system, as unfair as it seems, has one merit: it is relatively transparent. Many procedural documents can be viewed directly on the DOJ website or on www.pacer.gov. This has enabled me to build up a wealth of case law on the FCPA. Alstom's guilty plea is therefore accessible to all, and constitutes an extraordinary source of information, which very few French journalists have exploited. Let us start with the date on which this document was signed: 22 December 2014 (note that the most important provisions were signed on 19 December 2014, i.e. the same day as Alstom's extraordinary shareholders' meeting approving the takeover by GE). I do wonder why this agreement wasn't signed sooner. Six months earlier, in June 2014, negotiations were completed or in the process of being completed, so GE must have known the approximate amount of the fine, otherwise how could they possibly have determined an overall purchase price for Alstom, including the fine? Why then did the DOJ wait so long to close the deal?

In my opinion, there is only one plausible explanation: Patrick Kron had to be maintained in his position, to ensure that the shareholders would approve the sale. Indeed, if the guilty plea had been disclosed several weeks or months before the shareholders' meeting, its content could have caused an outcry and pushed Kron to resign. The Americans knew that Patrick Kron alone had a veritable personal interest in pushing this deal through. Let us go back to the content of the document. I observed that the prosecutors confined their prosecutions to just five countries. And yet I know all too well that the DOJ had the clout to access a considerable number of consultant contracts concluded by the group over a decade and on a global scale. The legal consequences could therefore have been much more serious. Here again, I suspect GE was wielding its influence, not wanting to name and shame all of Alstom's unscrupulous clients, as the latter would in turn become its own clients after the takeover.

It is also worth noting that the vast majority of the $75 million of bribes were paid after Kron became CEO of Alstom (in 2003). The last payments were made in 2011, as highlighted by the journalist Caroline Michel in *L'Obs*. This gives a finer understanding of the penalty Patrick Kron would have been exposed to had he been prosecuted. Using the same sentencing guidelines that the prosecutors used to calculate my sentence range (from fifteen to nineteen years in jail) for my involvement in the Indonesian case alone, I can't imagine what sentence he would have received for the whole shebang. He would obviously have ended up pleading guilty like me, but he would not have got away with a jail term of less than ten years. However, the DOJ chose to spare him. Meaning that three of the four individuals indicted, Rothschild, Pomponi and myself, were charged solely for the Tarahan project, which represented less than $600,000 in commissions out of a grand total of $75 million. Hoskins, the fourth person arrested, was indicted for a different Indonesian project. But when it came down to it, the prosecutors didn't go after anyone for the remaining $74 million. This clearly shows selective prosecution and that the intention of the US was not so much to punish the 'guilty' as to force Alstom's management to comply. At least Kron managed to avoid a string of charges. A handful of executives must undoubtedly be beholden to him. And some of them were likely to have had their share of the millions that were handed out upon signature of the deal with GE.

Upon reading Alstom's guilty plea, another 'untruth' by its management was exposed. In order to promote the best practices that he allegedly introduced, Patrick Kron had always boasted that no 'monitor' had been enforced on the company by the United States. The reality was quite different. Admittedly, it is usual for a guilty plea to contain a clause compelling the company that has admitted its wrongdoings to accept the presence of an external controller, i.e. a *monitor* on its premises for a three-year period. Such a monitor is normally a US lawyer responsible for verifying that the group thoroughly complies with its commitments to refrain from carrying out further acts of corruption. However, in Alstom's case this measure was unnecessary, since in December 2014 the group was already under monitoring by the World Bank, as part of its settlement in the Zambian corruption case.

Finally, I never cease to wonder about the role and responsibilities of Alstom's auditors in this fiasco. How could they have possibly overlooked $75 million in bribes? And, above all, why did they not insist on a reserve in Alstom's accounts for the fine that the company would have to pay? How could auditors approve accounts with a reserve of just a few tens of millions of euros, while the fine was estimated at around $1 billion? This strangely did not shock the AMF (the French Financial Markets Authority), which didn't investigate this cover-up of $772 million. At least to my knowledge. Finally, it is worth noting that, to date, Patrick Kron has not been prosecuted in France even though, by virtue of Alstom's guilty plea and especially the 'statement of facts', he has effectively admitted his culpability in one of the most extensive worldwide corruption scandals ever masterminded. In other circumstances, the PNF (French National Financial Prosecutor's Office) is quicker to initiate investigations (such as in early 2018 in the Bolloré case, where the sums at stake were much lower than in Alstom's case). We had to wait until 22 July 2019 to see the French anticorruption organization 'Anticor' file a complaint in Paris for corruption targeting without naming him, Patrick Kron.

'If the legal person Alstom has, by a guilty plea, acknowledged corruption that has been perpetuated for many years and if a scapegoat has had a long stay in an American prison, the physical persons guilty of the corruption have not been prosecuted, neither in France nor elsewhere,' says Anticor in a statement.

# 41

## French MPs confront Patrick Kron

It turned out to be a flash in the pan. Nothing more. And there was me thinking that the revelations of the American prosecutors at the end of December 2014 were going to create a huge media buzz in France. Alas, I was mistaken. A few articles touched on the subject, but they only mentioned one point: that Alstom had settled its US debt.

In the end, I was right to keep my mouth shut. I would have exposed myself to unnecessary risk by telling all. Especially since progress has stalled on my file in the United States. I'm deadlocked and cannot anticipate or plan anything. Sometimes I feel like I'm living in a state of permanent transit, like a passenger waiting for a plane that never arrives.

In the meantime, I decide it's best to be cautious and keep a low profile. It is the beginning of 2015 and I attend several dinners under the 'Chatham House' rule, whereby the content of the discussions is not disclosed outside. One such dinner was organized by Éric Denécé, the head of the CF2R think tank. I am his guest of honour. Around the table there are about twenty people, including two parliamentarians, senior government officials, a senior executive of the BNP, a former police commissioner turned intelligence officer, a former industrial tycoon who headed several multinationals, plus two investigative journalists. I also met with several leaders of France's CAC 40 companies. Tirelessly, I recount my experience and I alert them to the dangers, hoping to be heard.

Fortunately, some politicians are on the ball. They express their outrage, calling it an 'industrial scam'. There are not many such individuals. A small group of about forty members of parliament, the vast majority of them conservatives, joined forces to lobby twice, in June and December 2014, for the creation of a Parliamentary Enquiry Committee to investigate the sale of Alstom. Among those who supported this initiative were Henri Guaino, Jacques Myard and Philippe Houillon. But the most active of them was without doubt Daniel Fasquelle, the conservative MP from the Pas-de-Calais, who was party treasurer, law professor and former member of the Parliamentary Enquiry Committee into the Cahuzac case (Jérôme Cahuzac, a French politician found guilty of tax fraud). This initiative is hampered, however. The government opposes it, the Socialist Party does not budge, the Conservative party (UMP) abstains.

The forty rebel parliamentarians get a consolation prize, however. The Economic Affairs Committee (with more limited powers than the Assembly Committee) agreed to conduct a series of hearings on Alstom. The first debates were scheduled for 10 March 2015.

To be honest, I feared a fragmented debate, superficial questions, sterile arguments ... I have never really trusted politicians to deal with major economic issues. Well, I was wrong. The sessions proved to be fascinating, even if they were not always void of political considerations.

The first to enter the game was Fasquelle, who left his mark.

'I regret that our request for an investigation was not accepted. It is a pity, because the people interviewed in this context would have had to take an oath, which is not the case before this Commission.'

The Socialist President of the Commission, François Brottes, immediately dismissed his comment as being in contradiction with the views of his own political party.

'Each group has the right to request the creation of an enquiry committee. The UMP could use theirs if it wished.'

'Certainly,' replies Daniel Fasquelle, 'but you are proposing a series of interviews that remind me of a popular proverb: "Whoever hears only one bell hears only one sound."'

Proud of his words, the parliamentarian then proceeded with a series of very specific questions for Patrick Kron. The latter appeared before the MPs, accompanied by his faithful lieutenant, Poux-Guillaume, the same person who reached out to GE in the summer of 2013.

'Mr Kron, why such a hasty sale? The financial situation of Alstom, i.e. an order book of €51 billion, representing two and a half years of business, a turnover of €20 billion, an operating margin of 7 per cent driven by the activities of the energy division, net earnings of €556 million, does not in any way justify such haste.'

Then Fasquelle moved on to proceedings in America.

'What about the pressure exerted by the American judiciary on Alstom? We cannot help but compare this affair to GE's takeover in the past of other companies under investigation in the United States. Is there not part of a pattern that allows the US conglomerate to take over companies that have been severely weakened by legal proceedings? This question is highly significant, as it concerns not only Alstom, but other French companies too.'

And Fasquelle is not the only one to subscribe to this theory. On the left, too, and especially within the Communist Party, some MPs shared the same views.

'Mr Kron, we are dealing with a serious matter, which is nothing less than the carving up of one of our industrial hallmarks,' says André Chassaigne indignantly. 'This deal reflects the US strategy of economic domination, which is extremely serious for our national independence.'

As at the shareholders' meeting, Patrick Kron is not in any way disconcerted by this flurry of incriminating questions.

'I regret that I am unable to take the oath, but it doesn't change anything concerning the sincerity or transparency of my words.'

And then he embarks on a long explanation of his position.

'I consider the proposed merger between Alstom and General Electric to be beneficial for Alstom, beneficial for employment and beneficial for France. You may not agree, but it is clear that all the elements that led me to promote the sale are on the table. Mr Fasquelle, this operation was not carried out in a precipitous manner. It is exactly the opposite. Since my role is to anticipate, I have been searching for structural solutions for years that will enable Alstom to emerge from its difficulties. Do you really think that I did not first and foremost attempt to find a French solution allowing us to remain in the driver's seat? I could not find one and therefore I made the decision to approach General Electric. It was a carefully thought-out decision and we deliberately didn't publicize it from day one because, in our business, the slightest hint of financial difficulties can cause our clients to panic.'

He then once again denied any link between the US legal proceedings and his decision to sell. He even called it a plot against him.

'As to any conspiracy theories concerning the Department of Justice influencing this sale, you should be aware that an investigation into corruption was carried out against Alstom prior to our talks with General Electric. It is therefore insulting to say there may have been any collusion, and this is contrary to the facts.'

His argument does not convince me. To put it bluntly, it's ridiculous. Of course, the DOJ investigation was carried out prior to talks with GE, as it began in 2010. It was precisely because of the legal threat to himself and his company that Patrick Kron approached GE. Moreover, the members of parliament seem no more convinced by his idle explanations than I am.

'Are you speaking to us about a conspiracy theory to brush aside the corruption issue?' asks Clotilde Valter, former adviser to Lionel Jospin, a Socialist MP from Calvados. 'That's too easy. First, we need to examine the reasons why France lagged behind in prosecuting Alstom for corruption. How do you explain that Alstom was regularly and repeatedly under investigation by foreign authorities?'

'I repeat,' insists Patrick Kron, 'the Department of Justice's prosecution file is totally independent of the decision to sell the business.'

A touch annoyed, he reminds us that General Electric is to pay €12.35 billion and that therefore 'Alstom transport will be totally debt-free'. Finally, in front of the national delegation, Kron uses patriotism to reinforce his argument.

'I am a product of French meritocracy. My parents were immigrants. I am proud to say that I have hired almost fifteen thousand employees in France since I joined Alstom. We all make our contribution to employment in France and I also strive to make my modest contribution.'

And he pompously concludes by singing the praises of his deal.

'I repeat, I am proud of this transaction. You can consult all the journalists around the globe, you can conduct all the studies you want by any ministerial cabinet you want, you will see that there is not a single factor that contributed to the decision to promote this project that is not in the public arena. Anything else is insulting to me, slanderous and simply untrue. That's what I wanted to say to you. I may not be swearing on oath, but I'm looking you in the eye.'

Patrick Kron then leaves the chamber. I do not know if his mere glance was enough to convince the members of the parliament. Just one hour after his performance, though, his ears must have been burning and quite fiercely. The Minister of the Economy, Emmanuel Macron, in turn heard by the Commission, accused him outright of treason.

'Even though the state had initiated a strategic analysis of Alstom's future and demonstrated its willingness to work with the CEO and his shareholders, its CEO conducted a transaction behind the state's back that was not optimal in terms of national strategic interests. I repeat, we have all been presented with a *fait accompli.*'

And, according to the Minister, the consequences of the CEO's disloyalty was irreversible. 'We did not have enough time.'

Consequently, it was no longer possible to counter GE's offer. It was also unthinkable to forge an alliance with a major European group such as Siemens in such a short period of time, as Arnaud Montebourg had advocated, because, as Emmanuel Macron once again stated, 'taking longer would have been a major industrial mistake'. To sum up, the French government had been outperformed by a cynical industrial boss and it was too late to retract. If this explanation is accurate, then I find it tragic. The double act of a CEO was all it took to 'incapacitate' the government of the world's fifth-largest power. That's rather worrying to say the least. The sequel to Emmanuel Macron's performance proved to be even more pathetic. The Minister of the Economy is now questioned by the honourable member Fasquelle on the corruption cases. To my great amazement, just as I thought he was going to dodge the issue, he drops a real bombshell.

'With regard to the DOJ investigations, I asked Mr Kron this question very directly. Personally, I was convinced of the causal link between this investigation and Mr Kron's decision, but we have no evidence. Mr Kron has assured me that the US legal proceedings had no influence on his decision. I won't say that my personal conviction doesn't match yours, Mr Fasquelle, on some of your questions, but, I repeat, we have no way of proving it.'

I was stunned by this speech. Emmanuel Macron was 'convinced' that the US prosecution procedures were the reason for the sale, but he has no way of proving it. If the Minister of the Economy is not able to prove it, then who can? In any case, his Ministry was aware of the facts when I was there at the end of 2014, during my debriefings by the economic intelligence unit. And Claude Revel, the interministerial delegate for economic intelligence reporting directly to the Prime Minister, also knew, since she tried in vain to alert them. If the government knew the facts behind this sale, why didn't it block it? Or at least halt it until light had been shed on this sell-off. And why did Emmanuel Macron abandon Arnaud Montebourg mid-battle? Montebourg was the only one who tried to stop this suicide mission. How do you explain such collective resignation on the part of our political leaders? On several occasions, Matthieu Aron tried to obtain answers. Emmanuel Macron's cabinet did not wish to answer him and Michel Sapin, the then Minister of Finance, cautiously refrained from commenting. Arnaud Montebourg was the only one willing to discuss this pathetic episode. His explanation was simple, perhaps simplistic even.

'Because the Americans scare the hell out of them. They consider them too powerful,' he explained to Matthieu in an interview in June 2016.

Arnaud Montebourg will later declare in July 2019 during a hearing in the French Senate that 'Mr Kron has betrayed his country by selling (Alstom Power) in order to protect himself and to avoid prison for committing acts punishable under US law.'

Before the Economic Affairs Committee of the Assembly, it is now up to the unions to shed light on the sale of their company. Until that moment, they had kept in the background. Patrick Kron therefore had no problem arguing that he had their support. But on 10 March 2015, another fable of the Alstom case was shattered.

'This project is not an alliance, but a pure and simple sell-out,' exclaims Laurent Desgeorge, deputy coordinator of the union Inter-CFDT, who is concerned about the social consequences of the sale.

'GE may well be committed to the creation of thousands of jobs, but it will certainly not be sufficient to compensate for all the job losses over the next six years.'

The same tone was expressed by CGT representative Christian Garnier.

'We have been sold down the river; it is a pure sell-out of our energy sector to General Electric. There is no industrial strategy behind this, it's merely a financial political operation, and note that I have chosen my words carefully.'

Finally, Vincent Jozwiak, employee of Alstom Transport in Valenciennes, trade union representative, raises quite 'naturally the question of how the legal proceedings against certain Alstom officials influenced a small group of people, who decided under a veil of secrecy to sell off our energy activities to General Electric'.

Faced with this avalanche of criticism, the MPs unanimously decided to summon Patrick Kron once again, an approach that is quite rare in such a situation. This second hearing, which takes place on 1 April 2015, reveals nothing new however. Except for one point: the record bonus awarded to Patrick Kron by Alstom's Board of Directors. Once again, the honourable member Fasquelle goes in headfirst.

'At your previous hearing, you stated that the Board of Directors deemed that the agreement to sell to GE merited a bonus and decided to grant you an additional bonus of four million euros. I disagree with the position of your Board of Directors, and Mr Macron, Minister of Economy, Industry and Digital Affairs, considers that this bonus is "contrary to the ethical principles that large companies should uphold". Are you, like other business leaders, going to waive this bonus that the Minister deems unjustified?'

I shall never forget Mr Kron's reply to this question. If only on account of his incredible brazenness.

'I have absolutely no intention of forgoing the bonus of four million euros, because this would be a terrible loss for French taxpayers, who will benefit from a good portion of it, which is something you should be pleased about as representatives of their collective interest.'

In reality, Patrick Kron (who left Alstom at the end of 2015) hit a jackpot in excess of €4 million. Indeed, the Board of Directors awarded him for his final year of service within the company (financial year 2015–16) a fixed and variable compensation of €2.26 million. To this tidy sum, we can add his bonus of €4.45 million, as well as a top-hat pension plan. Alstom set aside a provision of €5.4 million with AXA the insurer, to cover the payment of an annual pension of €285,000. In total, it could therefore be estimated that the CEO would leave the company* with more than €12 million. It's quite simply outrageous. Due to his refusal for almost two years to negotiate with the DOJ, Kron is effectively the primary culprit behind the carving up of Alstom. Not to mention that he also put some of his senior executives at risk. Starting with me.

Others have done a better job of protecting their employees. For instance, in the Tarahan case, our Japanese consortium partner Marubeni was also prosecuted and had to plead guilty (it was fined $88 million), but none of its employees were arrested, let alone incarcerated. Yet the allegations against the Japanese company were identical to those against Alstom, as we had formed a 50/50 partnership. Furthermore, Marubeni had recruited and paid the same two consultants the same amounts as we had. The difference being that the Japanese firm admitted its culpability straight away and signed a guilty plea instead of dragging its feet like Alstom. Their strategy can be summed up as follows: 'OK, so you caught us red-handed. We admit our guilt, we'll pay up, but we are not opening the door any wider for you to investigate all of our global activities and blame us on other projects.' By adopting this strategy, they immediately limited the financial and logistical damage. This was

* In July 2016, the Alstom shareholders' meeting does, however, express its disapproval of this remuneration. More than 60 per cent of the shareholders disapproved of it. The Board of Directors then announced that it would re-consider Patrick Kron's compensation. Nevertheless, in November 2016 it endorsed it.

the complete reverse of the catastrophic tactics adopted by Patrick Kron. It is to be noted, however, that the DOJ showed surprising leniency towards Marubeni. Maybe they didn't want to take their investigations too far. Marubeni is a strategic partner of many American corporations operating on continents with a high corruption-perception index, such as Asia and Africa. In addition, the Tokyo-based Marubeni Corporation often worked in partnership with GE to compete for business (power plants or medical equipment).

# 42

## Final impediments to the sale

Despite being 'grilled' by national representatives, Patrick Kron received preferential treatment from the American judiciary in the spring of 2015. Strangely enough, he managed to obtain a grace period to pay the $772 million fine imposed on Alstom.

According to the DOJ's very stringent rules, the company plea agreement should be accepted by Judge Janet Bond Arterton through a formal sentencing. In FCPA cases, such sentencing occurs usually within one month of the guilty plea signature. Then the company has ten days to pay its fine. Since Alstom pleaded guilty on 22 December 2014, the sentencing should have occurred in January 2015 and the $772 million fine should have been paid by the end of the month. But, here, something unusual occurred. Alstom submitted to Judge Arterton a motion asking her to delay by six months her sentencing of the company (and therefore the payment of the fine) in order for them to allow time to complete the sale to GE. Strangely enough, this motion was unopposed by the DOJ. And even more strangely, Judge Arterton accepted this motion! Even the American press was surprised by this. The *Wall Street Journal*, in its edition of 1 February 2015, emphasized that 'The French group was given far better treatment than other companies'. The American daily even went as far as to question the judge in charge of the case, Janet Bond Arterton. The latter acknowledged that she had 'set a very leisurely timetable'. Three days later, on 4 February 2015, the *Wall Street Journal* continued to delve deeper and revealed – based on transcripts of the hearing of the French group's guilty plea – that GE's legal team, who were assisting Alstom, had been very closely involved in the entire negotiations with the DOJ.

In the newspaper's columns, Robert Luskin from Squire Patton Boggs, one of Alstom's lawyers, told a judge during a December 2014 plea hearing that 'General Electric reviewed the documents pertaining to settlement with the Department of Justice at all stages of their preparation and negotiation.'

This statement is quite astounding. Did this mean that the US giant, though not yet owner of Alstom, was able to access all the contracts signed with intermediaries over the last ten years? During an acquisition, such highly sensitive information is normally only reviewed once the sale has been finalized. If my supposition was correct, here, under the aegis of the DOJ, Alstom had disclosed irrefutable evidence to GE, its rival, of the widespread corruption system it had set up, including the names of its own managers involved and/or accused. Many of them had their work contract terminated during this period.

Highly embarrassed – I presume – by Alstom's lawyer's statement, the DOJ was forced to explain itself: 'The sale to GE was not a key factor in the government's resolution,' says Leslie Caldwell, head of the DOJ's anti-corruption unit.[*] Duly noted, but although it was not 'a key factor', it implies that the sale nevertheless had some influence. This is, to my mind, a mammoth confession. It explains the leniency shown by Judge Arterton in granting Alstom a leisurely payment deadline, since in this case, the timing was crucial.

In fact, the request by Alstom to postpone its sentencing, the non-opposition of the DOJ to such an unusual request, the intervention of GE throughout this process, all this can be easily explained and understood if we consider that there was one last

---

[*] *Wall Street Journal* of 4 February 2015.

hurdle to be overcome before the sale of Alstom could be sealed: the approval of the deal by the European Commission.

In 2001, GE had its merger with the company Honeywell blocked by Brussels. It is out of the question that this sale should be obstructed by Brussels. This time, nothing is left to chance. The first requirement is that the now 'ever faithful' Patrick Kron remains in place, kept under pressure, as well as the French government, so that they all commit to this ultimate challenge and battle on behalf of GE to get the approval in Brussels. To achieve this, there is nothing better than to keep a legal threat hanging over Alstom and its CEO by not bringing the FCPA case to a close until the European Commission has endorsed the sale. And this is exactly what Judge Arterton's ruling allowed time and space for.

With the DOJ's blessing, there was time for the European Union Member States to approve the sale before she approved the guilty plea. The legal association between the two subjects was therefore indisputably established, contrary to Patrick Kron's assertion.

This time, GE really needed a helping hand. It is not an easy feat to obtain the assent of the European Commission. In fact, Europe gasped in amazement. On 28 February 2015, the European Commission launched an in-depth investigation. The Brussels experts expressed concern about the consequences on the European energy market, especially in the high-power gas turbine sector. Even before the sale, GE was already the number-one manufacturer of this type of equipment and Alstom the third-largest global player. Once the French company has been taken over, the American giant will have a quasi-monopoly in Europe, with only Siemens as a credible competitor.

'Such a concentration could be detrimental to innovation and drive up market prices for a technology that is essential to fight climate change,' warns the European Commission. So, to placate the Europeans, Jeff Immelt, CEO of GE, makes some concessions.

He agrees to divest certain assets, including part of his portfolio of power plant maintenance contracts, to a more modest contender, the Italian company Ansaldo. This move is designed to remove some of GE's market dominance, thereby coaxing Brussels to agree. The negotiations are, however, neck and neck. On 5 May, Jeff Immelt went in person before the European Commission, to plead his case and accelerate the approval process. This was in vain. Brussels deemed that it had not received all the necessary information and announced on 12 May 2015 that it was deferring its decision to 21 August 2015. Siemens was not totally out of the picture at this stage and exploited the argument of too high a concentration to lobby its case. Ultimately, it was France that saved the day. Emmanuel Macron, visiting GE's Belfort plant, sent a strong message to Brussels by publicly defending the American takeover of Alstom. The French government wanted to close this 'unpleasant' file. The deal must not be scuppered. The sale must go through to avert proceedings being reopened by the DOJ. If they end up indicting one of our biggest industrial chiefs, the impact would be disastrous. Macron pleads his case to Brussels for Europe to validate the sell-off of a flagship French industrial group to the American giant. What an extraordinary reversal of roles.

Then finally, on 8 September 2015, GE obtains the long-awaited green light. Alstom was also involved in these negotiations. To compensate for the transfer of a part of Alstom's activities to the Italian company Ansaldo, Patrick Kron even agreed to lower the sale price by a further €300 million. Meaning even less in the group's coffers. The last obstacle to the sale had been overcome. Nothing could stop it now. It was concluded on 2 November 2015. In the newspaper *Les Échos*, Jeff Immelt welcomed a 'strategic' acquisition, referring to it as a 'once in a generation opportunity'. France has parted with a paragon and returned with empty pockets.

On 13 November 2015, Judge Arterton finally sentenced Alstom by approving the guilty plea negotiated between the DOJ and

Alstom, which had been signed eleven months earlier. This was a first in the history of the FCPA. Patrick Kron can relax now. He has saved his own skin. GE is now at the helm.

The first visible consequence of this takeover was the announcement by the US giant to the French unions of a massive restructuring operation. Ten thousand of the 65,000 jobs in Alstom Energy worldwide were to be slashed. Europe was hit particularly hard, with 6,500 job cuts planned. Germany was the most affected, losing 1,700 posts, then Switzerland with a loss of 1,200 positions, then France. Eight hundred jobs are at risk in France. In April 2016, an estimated 2,000–3,000 Alstom European employees expressed their anger at a rally held in Paris. Their message was clearly displayed on banners, written in English, German, Italian and Spanish. The employees of the former Alstom felt betrayed. 'The announcement of the redundancy plan,' confided one employee, 'was a crushing blow. I wasn't expecting the job losses to be on such a large scale. They lied to us.' France, however, has been spared to some extent. Jeff Immelt gave assurances that he would compensate for the loss of jobs in France. He announced the opening of a digital software research centre in Paris, promising the creation of 250 jobs in 'services such as finance or human resources through a leadership programme for young graduates' and stated that the Belfort site would accommodate a shared services centre employing bilingual and trilingual staff. The extent of this is all rather vague.

In spring 2018, the US group had to acknowledge that, contrary to the pledges made to François Hollande, it had failed to succeed in creating a thousand new jobs. Personally, I am not at all surprised. It was evident that the merger of Alstom and GE was going to lead to a large-scale destruction of jobs, notably in support roles such as IT, HR, legal and so forth.

The honeymoon period was short lived. On 13 May 2016, Alstom Transport (the part of Alstom that remained) filed a complaint in the United States against General Electric. The French felt that the wool had been pulled over their eyes. At the time of the sale of Alstom's energy business, the Americans had agreed to sell their train-signalling business, as an offset. GE, however, started to drag its feet over this and appeared reluctant to proceed. The US corporation no longer agreed to the sale price. While the agreement that was initially signed provided for the use of a French law firm (to establish a definitive price), GE then decided it wanted the International Chamber of Commerce (ICC) to act as arbitrator. Alstom Transport was therefore forced to take the case to a US court to have its rights reinstated. This was the first hitch in their union.

General Electric was also in litigation with another major French player: EDF. The issue at the heart of their dispute can hardly be deemed trivial, given that the maintenance of France's nuclear energy plants was at stake. Since it had taken control of Alstom, GE inherited responsibility for the maintenance of the fifty-eight turbines that run our nuclear reactors. However, the world's leading power equipment supplier wanted to review the terms of this contract, limiting notably its financial liability in the event of an incident, as well as increasing the prices of spare parts. GE even went as far as suspending its services for a few days in February 2016 to exert pressure on the French group. Jean-Bernard Levy, CEO of EDF, wrote to Jeff Immelt to express his anger: 'EDF has been forced to implement emergency measures exceeding our usual back-up plan. This attitude, from a partner with which we have a long-established business relationship, is simply unacceptable.' Unimpressed by this letter, the director of General Electric Power strongly urged EDF to accept these conditions before 15 June 2016. The French electricity giant was furious at this reaction and threatened to suspend its commercial relations with GE in retaliation. The case seemingly stalled at this stage. Both partners put down their weapons, but for how long? By effectively taking control of all France's nuclear power plants, GE and therefore the US government now have a massive deterrent device at their disposal for future use. What happens tomorrow if France opposes the United States on a major international political issue? A similar situation occurred earlier, in 2003, when France refused to participate in the Iraq war. In a documentary entitled *Phantom War* (devoted to the Alstom affair), General Henri Bentégeat, former Chief of Staff of the French Army (2002–06), reports how the United States decided at the time to cease supplying spare parts to the French army. 'If the affair had persisted,' said the general, 'it could have rendered our aircraft carrier *Charles de Gaulle* non-operational.'

In the middle of 2016, my legal situation was still just as blurred. The date of my sentencing in the United States was being constantly pushed back. Under such circumstances, it is very difficult to have any peace of mind. I also had to deal with my litigation with Alstom before the French employment tribunal.

I am contesting termination of my employment contract on grounds of 'job abandonment'. In addition, my former company, which has displayed no goodwill towards me, has 'forgotten' to pay me some €90,000 in full and final settlement. I therefore decide to sue them.

# 43

## The employment tribunal

I can't believe what I'm hearing. For the first time, a magistrate is moved by my plight. The Adviser of the Employment Chamber of the Versailles Court of Appeal charged with ruling on the 'balance of my final pay' (i.e. my wages owed at termination) that Alstom owes me is shocked by the way my company has treated me. At the end of the hearing, before making her decision, she asks if I will agree to mediation. I say yes and then, two days later, Alstom also gives its consent.

We meet for a first session. Apart from the mediator and myself, among those present are Markus Asshoff my French lawyer, Alstom's counsel, and a lawyer from Brussels representing GE. That same day, General Electric had just taken over leadership of Alstom.

'We sympathize with what has happened to you, Mr Pierucci, and we wish to reach an out-of-court settlement,' she says in her opening remarks.

Out of court! That's easily said. Does she need reminding of the ordeal I have been through?

'You realize that I have served fourteen months in a maximum-security US jail and that I was only released once the French government had agreed to sell Alstom Power and Grid to GE.'

'If your government had not caused so many problems, you would have been released earlier,' she immediately retorts.

I am speechless. I truly had not expected such candour. This lawyer has just acknowledged in front of four people, including two counsel, that there is indeed a clear link between my detention and the purchase of Alstom by GE. In short, she has just admitted that I was used as an economic hostage.

At least she's honest. But once we start talking finances, her tone changes. GE's counsel digs her heels in, maintaining that Alstom owes me nothing.

Then, as if it were normal, she tells me that her employer, GE, will have to inform the DOJ prosecutors of the outcome of our mediation in any case. I can't believe it. What right does the US Department of Justice have to interfere in a civil procedure concerning a French employee, subject to a French employment contract governed by the French Labour Code with a French company and before a French court? The GE counsel sees no incompatibility in this whatsoever.

'In any case, my company won't act without the DOJ's agreement,' she tells us.

And indeed, while we end up meeting several times, she each time reports back to GE, which systematically has to obtain DOJ approval. At the end of the third meeting, she deigned to offer me €30,000 of the €90,000 claimed in full and final settlement, and even added that it was just out of 'kindness', because her company considered that it 'did not owe me a penny'.

I didn't want her charity and refused it outright.

Good job I did, as one month later, the Versailles court acknowledged Alstom's wrongdoings and awarded me €45,000, indicating that for the other €45,000, this should be decided as part of my larger case against GE when the Labour Court in Nanterre rules on the whole of my unfair dismissal case. However, as of August 2019, such a ruling in Nanterre had not yet occurred.

Before entering this litigation, my lawyers had of course contacted Alstom to try to come to an arrangement. Several meetings took place, the first of which had happened in the spring of 2015. The human resource director of Alstom came in person, accompanied by his counsel. I was assisted by my two lawyers,

Paul-Albert Iweins and Markus Asshoff. I offloaded everything at the outset: Keith Carr's reassuring words before my trip to the United States, the total lack of support after my arrest, their refusal to receive my wife at the Paris headquarters, termination of my employment on grounds of 'job abandonment', their cessation of payments to my lawyers, their pettiness with respect to calculation of the full and final settlement due to me, etc. etc. Above all, I wanted the HR director to send a very clear message to Patrick Kron. I wanted him to know that I was well aware of the ploys he was using to extricate himself from a legal bind and that I wasn't going to let him push me around indefinitely. I then left the meeting, leaving my lawyers to negotiate.

The stakes were high. I have suffered irreparable damage. At forty-seven years old I will never again hold a position equivalent to the one I held at Alstom. Given my criminal record, I don't even think I will be able to land any salaried position.

Surprisingly enough, at that time, my former company seemed prepared to listen to me. After several meetings, we even agreed on a compensation range and agreed to submit it to an independent arbitrator. It was in my interests to get the deal signed as soon as possible. A few weeks, or months later, GE would be closing the sale, and the few people at Alstom who might want to redeem their consciences by providing me with some financial compensation would no longer be in the company's service. We therefore prepared our submissions for a ruling at the end of June/beginning of July 2015. And then nothing happened. We received no news until the middle of September. Until the day the human resource director told us he was leaving Alstom, that there was no longer any question of arbitration. But to settle my case against Alstom, the company was prepared to sign a compensation agreement with me including a payment of a few hundred thousand euros. It was 'take it or leave it', but I had to give my answer there and then. Though this amount was far from being negligible, it remained much lower than even the base range we had set for ourselves. After tax, this compensation would just cover my American and French legal expenses, my travel to the United States, and the possible fine I would have to pay to the DOJ. The HR director knew this too.

How do you explain his sudden U-turn? I can only see one possible explanation. At the beginning of September 2015, the European Commission authorized GE's takeover of Alstom. The deal was done, nothing could stop it now. Thereafter, the Americans knew that they were in a strong position and no longer needed to use kid gloves. I therefore declined their offer. The human resource director left Alstom at the end of October 2015 and Patrick Kron left a few weeks later. The talks broke down.

My ex-employer had failed to compensate me fittingly, and my perception is that Alstom sacrificed me and then stabbed me in the back. Twice in fact. First and foremost, it had knowingly sent me into the lions' den, without warning me of the dangers. Then, after my arrest, it abandoned me and left me for dead, like a wounded soldier on the battlefield. But the crazy thing is that it could have turned out differently, although it took me some time to understand this. It was after Lawrence Hoskins' indictment (concerning his time as senior vice-president for the Asia region) that I began to ask myself a lot of questions. I was surprised to see how many millions he was spending on legal fees to his lawyers Clifford Chance to fight his case against the DOJ, after having already posted a $1.5 million bond package for his bail. Where was all this money coming from? I knew he was wealthy but not to that degree. I then learned that his legal fees were in fact covered by an insurance policy, and realized, amazed and horrified, that as a director I could have also benefited from the same insurance coverage to pay my legal fees.

Alstom had taken out a specific Directors and Officers insurance coverage (D&O insurance) to protect its senior executives. Yet, rather bizarrely, my company did not activate such insurance at the time of my arrest. It is simply outrageous that they failed to activate the insurance protection that covered me. This type of cover is intended to avoid conflicts of interest between an employee and his or her employer. In this way, the employee can have complete trust that their counsel is independent and not influenced in any way by the interests of their employer.

With hindsight, it makes a lot of sense to me, but when I was arrested on 14 April 2013, and in the weeks and months that followed, the possibility of this insurance, which could only have been activated by Alstom in any case, did not even occur to me. Why did Keith Carr not enforce my rights under this protection? And why did he ask the same law firm who represented Alstom (Patton Boggs) to choose and pay a counsel to represent me, thereby voluntarily creating a huge potential conflict of interest? It is almost as if my company wanted to keep me under control.

What's more, these contracts covering senior executives include a specific clause for indictments in the United States. The insurers know very well that the individuals prosecuted are almost all forced to plead guilty. Their legal fees are therefore borne by the insurer, even after they have admitted their culpability.

In February 2017, still seeking further information, I decided to attend the World Congress of Insurers in Deauville. While there, I had the opportunity to meet a manager from Liberty, Alstom's insurer. My contact was fully aware of the situation, and for good reason: the Hoskins file had already cost him $3 million in legal fees, he told me. He then confirmed that I was indeed covered, but that Alstom had never asked him to activate my cover. If this is true, then I still have the possibility of obtaining cover, by asking my former employer to activate the insurance policy. Once back from Normandy, I send an official letter to Mr Poupart-Lafarge, the new head of Alstom, requesting him to finally trigger that insurance cover. I also send a letter to the firm's general counsel, as well as to GE. All my correspondence falls on deaf ears since I never received any answer.

# 44

## Relentless blackmail

Not only have they robbed me of my existence, they are now tarrying in giving it back to me. They want to gag me for as long as possible. If they can delay my sentencing for another few years, they will succeed, and then nobody will be interested by my 'revelations' on the Alstom affair. It is the end of summer 2016 and I have been back in France for almost two years now. Two years in limbo. In these two years I have made four trips to the United States to fix a date for my sentencing. These four trips all proved to be a waste of time, since on each occasion I learned that Lawrence Hoskins' trial had been postponed.

Since Judge Arterton has upheld some of Hoskins' lawyers' arguments and dismissed part of the charges brought against him, the case may end up in the hands of the Supreme Court judges. That would be disastrous for me. It would mean I would be 'pending' for another two, three, or even five years. It's unthinkable! I'm going to crack. I can only see one way out. I must demand that they sentence me in the hope that the judge will understand the terrible predicament I am in. I risk going back to prison for a long time, but I'm willing to take the risk. I'm going to play my last card. On 1 September 2016, I therefore ask Stan to file my request for sentencing.

Three months have passed and I am desperate. In the meantime, Stan finally agrees with the prosecutor's arguments and, believing it to be in my own interest, withdraws the request for sentencing. I only discover this in December 2016. I feel deceived. I touch rock bottom. I have lost all my faith in him but cannot afford to pay another lawyer. I can no longer see any light at the end of the tunnel. All this puts a huge strain on my relationship with Clara. We don't see eye to eye on anything. This nightmarish situation has alienated us, and we argue non-stop.

To try to maintain some kind of balance, I submerge myself in work, conferences, business dinners. I even assist the economist Claude Rochet in organizing a half-day symposium at the Assemblée Nationale in November 2015. The title of the conference couldn't have been more explicit: 'After Alstom, who is next?' I continue to provide assistance to companies and am swamped with requests in France and abroad. I also give talks (in confidential circles) in Spain, the UK, Poland, Germany, Belgium, Slovakia, Sweden, Switzerland and Holland. These conferences prove to be a huge success and, slowly but surely, my consultancy firm, which advises companies and their managers on the US Foreign Corrupt Practices Act and on procedures to be implemented for their own protection, takes off. I am not yet able to draw a salary, but its performance is pretty good.

My unique experience is particularly sought after in France, as there is a growing awareness here among business leaders of the urgent need to implement and comply with anti-corruption measures.

In December 2016, a new anti-corruption law, known as the Sapin II law named after the Socialist Finance Minister Michel Sapin, was promulgated in the Official Journal. It requires all French companies with a turnover of more than €100 million and more than 500 employees to implement anti-corruption processes, modelled on the UK Bribery Act and the US Foreign Corrupt Practices Act. It notably introduces the Judicial Convention of Public Interest (JCPI), which is directly inspired by and based on the deferred prosecution agreement. It is an agreement that allows a company to admit the facts without pleading guilty. The JCPI marks a minor revolution in our criminal procedure. A French anti-corruption agency was also created. This law, however incomplete it may be, was a first step towards protecting French companies from American or even British interference. It is unfortunate, though, that the Finance Minister, Michel Sapin, couldn't find a better way of presenting this new anti-corruption mechanism to businesses than to do it via a conference co-organized by a major American law firm in Paris and the France–USA Foundation. Couldn't he have announced the scoop to a French law firm? Another example of US omnipotence in all its splendour.

On the Alstom side, there are no new developments except that the scandal of the forced sale to GE is starting to make political waves. Candidates for the presidential elections even mentioned it in the first series of televised debates. Some of their teams contact me, but I keep my distance. I do not wish to be exploited or manipulated by them. I don't care whether they are right wing, left wing, centrist or extremist, this scandal transcends political divisions as our national security is at stake.

In my US prosecution file, things are going from bad to worse. The prosecuting attorneys are now suggesting delaying my sentencing until the autumn of 2017. Why wait so long? Are they trying to silence me during the election period?

Between the two rounds of the French presidential election, Marine Le Pen, who has been poorly advised and demonstrates poor knowledge of the Alstom case, gets her wires crossed when she tries to attack Emmanuel Macron on Alstom. Time is moving on. In May 2017 Emmanuel Macron is elected President. And in June 2017, his party La République en marche wins the parliamentary elections with flying colours. Then, in July 2017, I finally receive my summons. My sentencing hearing will be held on 25 September 2017.

I must now start work on my pre-sentencing report. The probation officer who drafts it must interview the defendant in order to gather his/her version of the facts to recommend a sentence to the judge, based on the corresponding sentencing guidelines, the risk of re-offending and the defendant's personal situation. All this appears reasonable on the surface and helps fuel the myth that the American justice system is fair. I was soon to be disappointed. While I finally see this as an opportunity to be able to explain the functions I had and my level of responsibility within Alstom, Stan advises me otherwise.

'If you do that, it'll alienate the prosecutors. The only thing that the probation officer wants to hear is that you are a good father, a good husband, a respected member of your community, and you go to church every Sunday.'

So be it. The telephone interview was only twenty minutes long and the probation officer did not ask me any questions about the Tarahan case or even about Alstom. To make my voice heard, all I have left now is my sentencing memo. I am going to ask for a 'time-served' verdict, which corresponds to the sentence that I have already served (i.e. the fourteen months in Wyatt). Stan approves. He thinks there is only a 'negligible' chance that I will be sent back to jail. But while everything seems to be on track, the situation once again spirals out of control in an unexpected way in a matter of days.

At the end of August 2017, Stan sends me an alarming message.

'We're in trouble. I have just received the prosecution's written submissions. We need to talk ASAP.'

When I read the document, I am furious and start to panic. The prosecutors have found new charges against me. First, they consider that I benefited from personal gain in this affair. Of course, they know that I did not receive a single kickback in any form whatsoever. But they have now decided to take into account the bonus that my employer paid me in the year in which the Tarahan contract was concluded. Like all other executives, I effectively received an annual bonus of up to 35 per cent of my salary. However, upon close examination, I was able to establish that the Tarahan deal only accounted for $700 in the computation of my bonus for that year. It is just absurd that such a ridiculously modest sum, which was part of my pay, should be held against me.

But that's not all. The worst is still to come. In their submissions, the prosecutors also recalculate the range of my prison sentence. They have just added four points (which must then be converted into years of detention), declaring me to be the 'leader' of the conspiracy. I am stunned. At no time was this accusation mentioned or even suggested in four years of proceedings. On the contrary, from the outset, Novick emphasized that I was just a link in the chain. How can he assert the contrary today?

'Because he needs a chief perpetrator,' says Stan.

Alstom has just paid the biggest fine ever inflicted in a corruption case in the entire history of the United States. It is therefore inconceivable, in the eyes of the prosecutors, to close this file without convicting a ringleader. Yet who have they got on their hit list? Rothschild?

'Impossible,' continues Stan. 'He negotiated a virtual amnesty, having probably cooperated with the DOJ.'

Pomponi? He is dead. Hoskins? It is not certain that he will even be tried. Kron? He has managed to escape the clutches of the DOJ. The only one left is Pierucci, who shall be the fall guy. Prosecutors will then be able to boast of having captured the 'leader of the conspiracy' and thus qualify for an attractive promotion. This also explains why they want me to take the blame for a new case, the Bahr II contract (a coal-fired boiler built in India). A deal made at a time when I had already left the Alstom Windsor office for two years and for which not even Alstom pleaded guilty. It is disgusting. There are no other words to express it. Morals and ethics obviously don't enter the game. So what leeway do I have, faced with such injustice? Zero. Either I accept the DOJ's humiliating conditions and pray that the hearing goes smoothly, or I don't attend the sentencing hearing and abscond.

The latter option would have terrible consequences. My two American friends who put up bail bond for me would lose their homes and I would be under an international arrest warrant. I have no choice other than to accept this fool's bargain. I therefore agree to return to the United States to be sentenced at the end of September.

# 45

## The verdict

On 25 September 2017, a few minutes before the opening of my sentencing hearing, while waiting in the court room I observe in amazement a huge painting hanging on the wall. A portrait more than three feet high of federal judge Janet Bond Arterton. A tall, slim, elegant woman of about seventy years of age, with an impenetrable gaze, who has the typical look of the bourgeois families you see on the east coast of the United States. Though Janet Bond Arterton (the same judge who allowed Alstom two payment terms in 2015) has been following my case for more than four years, I have never met her before.

Yet this person will decide my fate. I looked her up. All I know is that this former employment lawyer, appointed judge by Bill Clinton, who showed such leniency to my former employer, has a reputation of being as hard as nails.

I therefore fear the worst. I'm very afraid they will put me back in prison, despite Stan's assurance that David E. Novick was very satisfied with the sentencing memo we had filed. Of course it satisfied him. I had given in to all their draconian edicts and arm twisting.

At 10 a.m. sharp, Judge Janet Bond Arterton opens my sentencing hearing.

'Good morning. Please be seated Mr Pierucci. Have you read the Pre-sentence Report that the probation officer has prepared concerning you?'

'Yes.'

'And do you understand the contents?'

'Yes, Your Honour.'

'Have you had the opportunity to respond to those contents, either speaking with the probation officer or through your counsel?'

I would love to tell her that I challenge just about every line of this simple copy-and-paste of the prosecutors' submissions, that I do not accept the chief conspirator role they have assigned to me, nor do I accept my implication in a case in India that had nothing to do with me, and that I had never once benefited from a personal gain in any of these cases … But it is too late. If I go down that road, I'll be facing a ten-year sentence. This is how the trap closes. With a churned stomach I mutter: 'Yes, Your Honour.'

'Good, so let us take a look at your sentencing range.'

And Janet Bond Arterton adds up the points, like a grocer counting her end-of-day takings.

'For an offence of bribery, if not a public official, the base offence level is 12.

'For offences involving more than one bribe, a 2-level increase. The value of the benefit received in return for the bribes is the combination of the Tarahan project and the Bond 2 project margins, which provides an increase in the offence level of 20. Because the offence involved elected public officials with high-ranking decision-making authority, a level-4 increase is called for. There is also a level-4 increase for our role as an organizer or leader of a criminal action. Finally, Mr Pierucci having entered a guilty plea and accepted personal responsibility for the offence, is entitled to a level-2 reduction. Just as a matter of form. Will the government file a motion for the third point?'

'We do, Your Honour,' responds Attorney Novick.

'Well, that reduces the offence level by 3, leaving us with a level 29.'

'Thirty-nine,' says Novick.

'I meant 39, thank you. Mr Pierucci has no prior criminal convictions. He is therefore criminal history Category I. The imprisonment range is 262 to 327 months.'

I must once again contain my anger. By accepting, as Stan advised me, all the conditions required by the prosecution, I have automatically increased the theoretical length of my sentence. I now risk up to twenty-seven years in prison.

Stan, who has always encouraged me to 'lie low' before the DOJ, starts his pleading. I am worried. I fear a disaster. And disaster strikes. His performance is not convincing at all, sometimes stammering and hesitating over his words. He at no time addresses the case on its merits and only talks about the extreme conditions of my detention at Wyatt. He reels the whole thing off in precisely six minutes. I'm worth six measly minutes! I find this appalling. Novick follows him and scarcely takes longer.

'Mr Pierucci, of course, was not involved in the entirety of the conduct that Alstom was engaged in. And here, Your Honour, it is true that there was a culture of corruption at Alstom that is reflected in the corporate pleas.'

At least Novick acknowledges that I am not the only one responsible. If nothing else. However, this doesn't inspire the judge to show any leniency.

'Nevertheless, the acts carried out by Frédéric Pierucci are extremely serious. And, as the government has also pointed out, this culture of corruption was reflected in the actions of the company's leaders, who failed to respect their moral, ethical and legal obligations.'

It is then up to me to conclude, by reading a prepared text, in which I admit my guilt and ask my family and friends to forgive my behaviour. The 'debates' lasted only thirty-eight minutes, and my only 'exchange' with the federal judge was the reading out of my repentance speech. At no time did Judge Janet Bond Arterton question me before sentencing me. She takes a few minutes' recess alone to reflect on the sentence to be pronounced on me. She has been gone a good half-hour. During this endless wait I do not utter a single word to Stan. He knows that his performance was disastrous and that his 'strategy' to not contradict the prosecutors was self-defeating.

I turn to face my father, who insisted on being at my side throughout this ordeal. His English isn't too good. He probably didn't understand much of what went on. But was there really anything to understand? Tom, seated next to him, translates some snippets for him. He is livid. Forty minutes later, Judge Arterton returns and asks everyone to be seated so that she can pronounce her verdict. By this time, I had understood that I would be returning to prison. I just didn't know how long for. Janet Bond Arterton begins to read out her judgment.

'It is hard to hear Mr Pierucci explain that he loves his wife, his children, his family, but at no time did he think about the consequences on them of his actions.'

This speech on personal morals is promptly followed by a second one.

'Corrupt individuals who accept bribes divert their country's meagre resources. And in these nations, efforts to establish democracies are undermined by the actions of international businessmen. Frankly, the court was disheartened today that Mr Pierucci gave no recognition in his apologies to the serious consequences beyond the impact on his family, which has been eloquently set out in the letters that the Court received.'

So, according to the judge, I should have 'apologized' because corruption exists in Third World countries. This is the height of absurdity for a country like Indonesia, which Suharto led for decades with the support of the US government in exchange for military protection and access to natural resources for its leading companies, which has made it one of the most corrupt countries in the world. This judge fully embodies American hypocrisy in all its grandeur.

But this is not the moment to be indignant, as she is about to pronounce her verdict.

'A sentence must serve as a deterrent, both to the individual defendant, and for others in the world of international business seeking the dollars of Third World countries for their projects and their bottom line.

'Mr Pierucci, will you please stand. For all the reasons which I have outlined, you are remanded to the custody of the Attorney General of the United States for thirty months, minus your time served and any time for good behaviour. You must report to the detention centre notified to you by the Bureau of Prisons on 26 October at 12 noon.'

I am devastated! Just yesterday, Stan seemed so sure that I would not go back to prison that I ended up believing him. What a fool I was. If I take into account the time I have already served at Wyatt, and the time deducted for good behaviour, I will still have to serve an additional twelve months! This is a curse. My poor family, what have they done to deserve such a punishment?

I turn towards my father. My two friends, Linda and Tom, are already explaining the verdict to him. I try to console him. 'Don't worry Dad, I'll be okay. At least, this way, it'll all be over in twelve months and I can start a new life again.' He remains silent and just looks at me sadly. He is completely despondent.

As for me, I'm very angry, with everyone. With Stan, the prosecuting attorneys, the judge, the system, Alstom, Kron, and especially with myself. How could I have trusted the US justice system, believing I would come out of it okay? And now I must announce it to Clara.

While Stan is negotiating with the prosecutor for me to return to France before coming back to the United States to serve my sentence, I find a quiet area and I call Clara. Normally so strong, she breaks down.

The severity of my sentence is unprecedented. This is the first

FCPA case to be judged in Connecticut and Judge Arterton wanted to make an example of me. I am paying for all the others, including those at Alstom who have got off scot-free. The only positive factor in all this is that I am no longer in limbo. For the first time in four and a half years I know where I'm going. I am dreading going back to jail of course, but in twelve months' time, this nightmare will be over. I have to stay strong for me, for Clara, for Léa, Pierre, Gabriella, Raphaella and for all those who are behind me. I am exceedingly lucky in so far as I am not alone in this.

The same day I am sentenced, whether by fate or coincidence, Alstom Transport announces its takeover by Siemens disguised again as a merger among equals. The German giant will take over the transport branch, while GE had acquired the energy branch. I am not surprised, as this is what the analysts had predicted. Only Patrick Kron believed Alstom could survive by relying exclusively on its transport activity. If he really did have that conviction? But that was three years ago, and everyone has since forgotten.

# 46

## Separated again

Everything is happening with the speed of light. The DOJ agrees to let me return to France, but I have to be back in the United States on 12 October without fail (two weeks before my incarceration starts). This means I only have a few days in which to organize my year-long absence.

Just before taking the plane to Paris, I managed to speak with Jérôme Henry, who had since become Deputy Consul in New York. When he was stationed in Boston, he came to meet me in Wyatt. He was surprised to see me again, having thought that my case had long since been closed.

'This is the first time,' he says, 'that I have heard of a sentencing that occurred four years after a guilty plea; it is just outrageous.'

He advises me to file my transfer request as soon as possible, so that I can serve my prison sentence in France rather than in the United States. He even makes me fill out the forms in his office so that he can expedite them immediately to the Ministry of Justice in Paris.

'On the French side,' he told me, 'the agreement will be immediate, but the DOJ will have to agree to it, and there I'm afraid it may take a little longer.'

But he is confident. I satisfy all the conditions required for a transfer: my sentence cannot be appealed since I waived my right to appeal when I signed my guilty plea, and I have no ties to the United States.

'Therefore, in principle, there is no reason that the Americans should refuse,' he reassures me.

I just pray that he is right. If I can serve my sentence in France, I would obviously apply for parole, and according to my lawyers Markus Asshoff and Paul-Albert Iweins, I have a good chance of being released quickly, perhaps with electronic tagging. I wouldn't therefore be separated from my family. How else am I going to explain the situation to my two youngest daughters, Gabriella and Raphaella (twins who are now twelve years old)? I discussed this at length with Clara. Finally, we agreed to tell them that I must return to the United States for a period of approximately six months, to a 'camp' where they will not be able to visit me. We have decided not to mention the word prison. We decided to announce it to them in the presence of the two eldest, Peter and Léa, who could console them and, above all, play down the situation. This is one of the hardest things I have had to do. My speech is clumsy, my words clash, my voice shakes. I try to contain my emotion and tears, but it's so hard. Gabriella bursts into tears and Raphaella, more reserved, goes completely silent. Gabriella bombards me with questions. 'Will you be back for Christmas? And for our birthday in January? Who is going to take us to school now? What is a camp? Is it like a holiday camp, with activities? Can we Skype each other? Why can't we come and visit you? Will you have any friends there? What work are you doing now? Why won't you be able to return to the States after? I like Americans. And when I'm a Hollywood actress, will you still come and see me?' With the elder twins Pierre and Léa (nineteen years old), it is of course different.

In 2015, I had explained to them in detail what had happened to me. They are both very intelligent and had understood the situation better than I could have imagined. Once the youngest were in bed, we decided to all watch the documentary *Phantom War* on the sale of Alstom, which had just been broadcast on LCP, the parliamentary TV channel. It is a real pity that this documentary was not broadcast on a channel with a larger audience. The authors conducted a formidable enquiry. They did an excellent job of analyzing the DOJ's influence on the sale of Alstom to GE. They heavily criticised Patrick Kron and much of France's political elite, including Valls, Macron and Hollande. Not forgetting Sarkozy. According to them, former French president Nicolas Sarkozy's law practice (Claude & Sarkozy) allegedly worked for General Electric. Above all, the documentary emphasized my role as an 'economic hostage'.

For all those who watched this documentary, i.e. Pierre, Léa, Clara, Juliette and my various friends, even though they knew the story, it was a real eye-opener. From a professional point of view, I also have to organize affairs relating to my small business so that it can continue to operate during my year-long absence. My business partners agree to hold the fort for me. In the two weeks before I leave for the United States, I attend a meeting with about a hundred executives of a major blue-chip CAC 40 company. In the business world, more and more people are starting to figure out the hidden story behind Alstom. In the political world, too. A few months ago, a new delegation of the Assemblée Nationale led by MPs Karine Berger (socialist) and Pierre Lellouche (centre-right) examined the extraterritoriality of US legislation. The MPs, who travelled to the United States and met with DOJ and FBI officials, were alarmed by the scale of the phenomenon.

'If a company has so much as half a foot on US territory, the US authorities consider that it is subject to US law,' lamented Karine Berger.

Even more worryingly, the American authorities have acknowledged outrightly, in front of the French delegation, that they will not hesitate to 'deploy all the resources of the NSA [National Security Agency], the organization in charge of all electronic surveillance and wire-tapping, to open their investigations'. Finally, in addition to the list of convictions for FCPA offences, the parliamentary delegates drew up a list of the companies most heavily sanctioned for violating embargoes and money-laundering legisla-tion. It turned out that fourteen out of fifteen of them are European!*

Only one US company had been collared: J. P. Morgan.

In the two weeks left before I had to return to the United States, I also met two former government ministers on separate occasions. They both expressed concerns for my safety. They even advised me not to go back to the United States. They feared what might happen to me in prison. I think they are blowing it out of propor-tion. At least I hope so, because when you receive such a warning from former government officials, there is reason to be concerned. In any case, they promised to alert the Ministry of Foreign Affairs so that my transfer to France could be accelerated. One of them sent a very detailed report about me to Philippe Etienne, Emmanuel Macron's diplomatic adviser.

The date of my departure is fast approaching. Aside from my family, I say farewell to all my friends who have stood by me during these last few years, including Antoine, who visited me several times; my long-standing friend Leila; Didier and his wife Alexandra, who managed to raise my spirits each time they visited; Deniz, multilingual legal translator, who tirelessly assisted me in all my procedures; Leslie, Alexandre, Pierre, Éric, Claude, Claire and many others. A wave of anxiety sweeps over me just before boarding. Apparently, my special visa and accompanying letter are insuffi-cient. The security officer takes me aside after all the other passen-gers have embarked. He has received orders to call a special number in the United States, where the time is 5 a.m. He tried several times and finally got hold of his manager, who took nearly an hour to resolve the situation. My arrival in JFK is scheduled for 11 p.m., and the following day I must meet a probation officer at the Hartford court who will tell me in which prison I will be

* See Appendix 1.

incarcerated. On 23 October 2017, I receive the answer. I will be imprisoned in Moshannon Valley Correction Center (MVCC).

I look it up on the internet. It is hardly reassuring. Moshannon Valley Correctional Center is surrounded by gigantic swirls of barbed wire and is situated at an altitude of more than 3,000 feet on a desert plateau in the heart of Pennsylvania. Stroke of luck, however. One of my fellow cell mates at Wyatt, the Transporter, knows MVCC very well. He spent his last two years of detention there. He's going to tell his old buddies about my imminent arrival. I'm therefore going to be 'ordained' (admitted into their circle) and needn't fear for my safety. In prison your reputation is paramount, way more than outside.

On the morning of 26 October 2017, I order a taxi from State College, where I landed the previous evening from Hartford. We cross immense forests and the driver gets lost. Even with the aid of his GPS, it is difficult to find the prison. We finally arrive at the car park of the Moshannon Valley Correctional Center (MVCC). My driver kindly asks me at what time he should pick me up. I take his number and tell him I'll call him.

# 47

## My new prison

Unfortunately, it is all familiar to me. Same coloured walls, same furniture, same security checkpoint, same jargon, same smell, same humiliation . . . A step back in time. After the administrative formalities, I strip completely and, like in Wyatt, they give me three pairs of khaki trousers, three boxers and three T-shirts.

Moshannon Valley is home to 1,800 prisoners, all of whom are *aliens* (foreign nationals) who have less than ten years left to serve. The breakdown between nationalities is typical of this type of prison: about 900 Mexicans; 500 Dominicans; 200 'Blacks', mainly Nigerians, Ghanaians, Ivorians and Haitians; 50 Asians (Chinese, Indians, Pakistanis); 100 'other Hispanics' (Colombians, Cubans, Hondurans); and 100 'Internationals', a very broad category that includes all the others (Canadians, Europeans, Maghreb and Middle Eastern inhabitants). This correctional facility is managed by a private operator that has several other facilities of the same genre in the United States and abroad. Like any 'enterprise', GEO does its best to maximize its profits. The group therefore does not hesitate to cut back as much as it can on services such as meals, heating, maintenance of premises and medical assistance, or to increase the price of items that inmates purchase from the commissary, or to get the inmates' 'stay' extended to the maximum by sending them to the 'hole' (solitary confinement) so that they forgo part of their reduced sentence obtained for good behaviour.

MVCC has its own rules, but the inmates also have theirs. And these are not quite the same as those of Wyatt, as I'm about to find out. Here, only Mexicans, Dominicans, 'Blacks' and 'Internationals' have got rights. The others have none. Furthermore, the Mexicans and Dominicans enjoy more rights than all the others. They are the 'sheriffs' of the jail. They write their own laws.

On my first day of detention, they put me in C6 pod. This dormitory can hold forty-nine inmates, but they have packed in seventy-two of us! Beds have been added in every available space. There are no spare seats at the table. The 'Internationals' have the right to one table, the 'Blacks' two tables, the Dominicans four and the Mexicans six. The others are not allowed a seat at the table and must fend for themselves. And there is no question of trying to be generous and inviting someone to 'your' table; otherwise you risk being excluded from the pod.

Upon my arrival, thanks to the message that my former cell mate of Wyatt was able to pass on, I am lucky to be very warmly welcomed by 'Muay Thaï'. This is what they call the Slovak who spent five years in the Foreign Legion before becoming a mercenary. After battling it out during armed conflicts (Iraq, Sierra Leone, Congo, Yugoslavia), he settled in Thailand, married a local woman, opened a Thai boxing school, and trained many Mixed Martial Arts (MMA) champions, before being arrested and then extradited to the United States, where he was finally sentenced to ten years for a narcotics-trafficking case (a sting operation organized by the FBI). A second 'International', nicknamed Hollywood, a German national who introduced himself as the head of the 'Internationals' of Unit C, gave me a welcome package with the basic necessities until I received my first order at the Commissary (coffee, sugar, soap, powdered milk, dried mackerel in sachets). He is also a former mercenary. He was sentenced to ten years for being involved in a conspiracy to murder an American drug enforcement agent (here again, a sting operation orchestrated by the FBI). He helps me find my bed and settle in. Muay Thaï and Hollywood were charged in a case associated with that of the famous Viktor Bout, the Ukrainian arms dealer immortalized by Nicolas Cage in the film *Lord of War*. He is one of their buddies. I'm in good company here! In any case, they are very courteous to me and even offer me a place at their table. A privilege no longer enjoyed by the 'Internationals' in the pod next to ours, the C5. Their leader, a Bulgarian who was in high debt to the Mexicans, sold their table to the 'Blacks' for $400, at $100 a place. I had better learn the codes of this new establishment ASAP. Wyatt was a living hell, but now I'm wondering whether MVCC is even more grisly.

Whatever the prison, the hard reality is that time goes twice or even three times as slow as it does outside. So if I want to be back in France for Christmas, in two months' time, I had better speed up my transfer request. I approach the prison social worker, Ms H. To be able to be transferred to my country, I must obtain consent from the MVCC administration and the DOJ. A simple procedure in theory that proves much more complicated in reality. Things go wrong yet again.

On 28 October 2017, Ms H. summons me.

'I'm sorry, but I cannot finalize your transfer request. The bilateral agreement with France stipulates that you must have at least twelve months of sentence left to serve at the time of the request.'

'Yes, I know this. But since I was sentenced to thirty months, and I have already served fourteen months, I have sixteen months left.'

'No, if you deduct time for "good behaviour", you are left with less.'

'But you cannot calculate it in this way, as I haven't yet accumulated any time for good behaviour. It's just theoretical.'

'But that's how it's calculated. I cannot go against the process!'

She has just pronounced the fateful word. *Process!*

From this moment on, I know there is no point in insisting.

Fortunately, I managed to get in touch with Marie-Laurence Navarri, the French liaison officer based in Washington, who promises to intercede on my behalf. Whatever she said worked, as on 8 November 2017 I was once again summoned by Ms H., who this time is accompanied by her boss, M.J.

'That's correct,' he said to me. 'There was an error in our first calculation. Your release date, even with good behaviour, is 31 October 2018.'

'So you can now transmit my request for transfer?'

'Err no, it is impossible. Today is 8 November, so you're no longer within the one-year timeframe.'

'But at the time I applied I was in the right window. If I no longer am, it's because you made a mistake in the calculation!'

'Maybe, but that doesn't change anything.'

It's like talking to a brick wall. If I continue, I'll end up in the hole. There is nothing I can do. Meanwhile, I call back the French liaison officer, who is flabbergasted. She immediately grabs her phone and contacts one of the DOJ officials, who happens to agree with me. The MVCC administration, furious at being repudiated, drags its feet yet again, meaning my transfer request file is eventually sent to the DOJ on 6 December, more than a month and a half after I entered Moshannon Valley. I am going to have to spend Christmas in jail with my cell mates for company.

At least I know I can count on the unwavering support of Marie-Laurence Navarri, who promptly comes to visit me in prison. She tells me that the French Ambassador himself wrote a letter to Jeff Sessions, the United States Attorney General, to inform him of the French government's wish to see me return as soon as possible. However, she urges me to be cautious: 'Don't get your hopes up too high,' she said. And then she explains to me that other FCPA files concerning large French enterprises are currently under examination by the DOJ. Relations between Paris and Washington have thus become tenser. In addition, she tells me, there is the Committee.

'Which committee?'

'The Assemblée Nationale,' Marie-Laurence Navarri explains to me, 'has just opened an investigation into Alstom and more generally into US meddling. And its president, Olivier Marleix [centre-right], has every intention of questioning all witnesses, including Patrick Kron, under oath this time.'

At long last! I have been fighting for this for the past three years. Though, at the same time, it couldn't have come at a worse moment for me. In this context, I can't see the DOJ validating my transfer file rapidly.

# 48

## Violence and trafficking

My first impression was the right one: Moshannon Valley Prison is not as dangerous as Wyatt detention facility but the relationships between inmates are more vicious.

Sure, I was surrounded by hardened criminals at Wyatt and here the facility only takes in foreign inmates at the end of their sentences who are to be deported from the United States. Theoretically speaking, this implies inmates who are calmer than those in Rhode Island. However, given that security measures are less severe here, Mexican and Dominican gangs have seized power. They run a full-fledged clandestine economy, a kind of mafia.

It's very simple: at MVCC everything can be bought, and everything can be rented. This ranges from narcotics trafficking to gym 'places' ($5 a week for a one-hour slot a day). There are hairdressers ($2 per haircut), grocers (notably a Mexican who has accumulated an impressive amount of products, sometimes stolen from the kitchen, and sells them for 20 per cent above the usual rates), tattoo artists, electronics technicians (who repair damaged radios), cleaners (to whom pod cleaning can be outsourced), and finally a few prostitutes (prisoners who sell themselves for survival). There is also an impressive pornographic magazine business. Such magazines are so rare that they are worth several hundred dollars. Although prohibited, betting (on basketball or American football results) and poker are also a considerable source of income, and many prisoners prefer to go to the hole (solitary confinement) to avoid paying their debts. Telephone minutes can also be purchased from those who need money. The exchange currency is the *mack* (a bag of mackerel worth one dollar).

The MVCC administration prefers to turn a blind eye to all this wheeling and dealing. It literally exploits prisoners, whom it uses for the preparation and service of meals, kitchen cleaning, maintenance of buildings (painting, plumbing, rubbish collection …) and green spaces, and programme management (courses, bookshop …). All inmates are forced to work, from one to five or more hours a day. For the first three months of detention there is no choice, everyone is assigned to the kitchens.

Remuneration varies according to tasks and qualifications and is set at between 12 to 40 cents per hour. So, for my first month of work in the kitchen, where I am assigned to the dishwasher (five hours a day, three days a week), I was royally paid $11.26 (€9.80). And there's no way to escape it. This system is similar to modern-day slavery. Private investors have their 'Made in America' products manufactured inside this type of jail, where the labour costs cannot be beaten.

But there is even more hypocrisy to come. As we don't reside in the United States, we all have the status of *illegal alien* in the eyes of the US government. Besides, many MVCC prisoners have been convicted of re-entry (prohibited entry into the United States). These illegal aliens, after being deported a first time, were arrested for trying their luck a second time. These men (who cannot legitimately work in the US) find themselves forced to work within these four walls for a pittance, and this in total legality. The prison administration has in effect applied the famous Thirteenth Amendment to the American Constitution abolishing slavery 'except as a punishment for a crime whereof the party shall have been duly convicted'. So, legally speaking, we are all slaves. And woe betide those who refuse to bow to this discipline. They go straight to the 'hole' and are then transferred to other GEO group prisons.

The most recalcitrant are subject to a special programme known as *diesel therapy*. They are rotated every two or three days from one prison to another, constantly being transported from one end of the country to the other in minivans. It supposedly calms them down. In Texas, inmates recently rebelled against this system in a GEO facility. It was closed after being partially destroyed by fire during riots.

In Moshannon Valley, the most daring way to express dissatisfaction is to go on a 'counting strike'. They count us five times a day. At that moment, we must all stand silently by our beds. Two guards take turns to count us and write the result on a sheet of paper. If they come up with the same number, they proudly brandish their paper and express their satisfaction with huge grins. If not, they begin again. The 'strike' action therefore consists of being constantly on the move in the pod to prevent them from achieving their goal. Everyone is of course forced to join in the game and everyone does, at risk of being seen as a snitch.

To survive in this 'parallel world', I decided to continue writing my story. I send fresh scripts to Matthieu Aron, who edits and works on it. I also receive many letters from family and friends and reply promptly. And I start playing chess again, but the competition is very tough in the pod, where we have some excellent players. One of them, Chuck, a Hells Angels veteran sentenced to twenty-four years, who is due to be released next year, is just unbeatable. At the end of November, another good player arrived, a Briton nicknamed Fifa. He was arrested in Zurich just before the General Meeting of FIFA (International Federation of Association Football) in May 2015 and spent almost a year in prison in Switzerland before being extradited to the United States. Very quickly we strike up a friendship and discuss the similarities between our respective cases. According to him, the FIFA scandal (the payment of bribes to obtain the awarding of sports competitions) was nothing but revenge by the United States, furious at having been ousted by Qatar for the hosting of the 2022 World Cup. He is adamant, even though he does not give me any detail, that the Americans, despite taking the moral high ground, conduct themselves like most other countries. They also do not hesitate to lobby different federations.

Hang on in there and above all make sure you don't get punished and lose your good-behaviour points for sentence reduction. It's an everyday challenge. For example, when you work in the kitchen, you must steal food and bring it back to your pod. It's an obligation. If you don't, the other inmates will come after you. But if you get caught, you're sent straight to the hole, plus you face cancellation of your phone rights, and you lose twenty-seven days of sentence reduction for good behaviour. This just happened to a Mexican for a chicken leg.

I'm permanently on my guard. I have set myself a roadmap, a kind of checklist that I try to respect rigorously. Get into a routine, stay physically fit, don't look for trouble, don't gamble, don't accumulate debt, keep your head down, never complain, never brag, never lie about who you are or were outside, don't snitch if someone breaks the rules, never raise your voice, never get angry, never touch or brush another inmate, don't even talk to known snitches and convicted paedophiles, do not sit with inmates from another group, use your knowledge to help others but do not overdo it, create alliances, do not accept gifts that make you accountable, do not get involved in other people's affairs, do not ask to change TV channels (one of the main causes of strife), do not stare at anyone, do not pity others, and, most importantly, be patient.

On 6 January my eldest twins Pierre and Léa celebrate their twentieth birthday and I am gutted about not being with them. Then, on 14 January, I celebrate my fiftieth birthday. Filippo, the Greek who shared my cell during my last month in Wyatt and whom I reconnected with here, made two cakes that we share with the 'Internationals' of Unit C, namely Muay Thaï, Hollywood,

Vlad, two other Russians, two Georgians, one Romanian, and Fifa.

On 15 January I get some bad news The liaison officer, Marie-Laurence Navarri, has just informed my sister Juliette that the DOJ has refused my transfer to France. However, Ms Navarri does not give up. Upon the request of the French presidential office, which seems determined to help me, she drafted a letter that Emmanuel Macron was to send in person to Donald Trump, requesting a presidential pardon. I don't have much faith in it happening, but nevertheless cling to it like a last lifeline.

On 22 January it's the birthday of my youngest twins, Gabriella and Raphaella. I manage to speak to them for a few minutes by phone.

'Daddy? When are you coming back?'

They haven't asked me that for a very long time. All the bad memories come flooding back.

'I don't know Gabriella, but soon.'

'You always say that. Like before the last vacation. I heard you and Mum talking about Emmanuel Macron. Is it up to him?'

'It's complicated, but yes, a little. You must be patient a bit longer sweetheart.'

'If you don't come home, I'm going to write to Emmanuel Macron that Daddy has to come home. And I will go on strike at school with all my friends.'

After hanging up, I suddenly feel very depressed. It doesn't happen to me very often. The last time was in Wyatt when I heard that Alstom had fired me. As a prisoner, you cannot and must not confide in anyone during such moments. They'd take you for a loser, a wimp, a wuss. So you grit your teeth, say nothing and continue to act as if everything is fine. But damn, it's hard!

# 49

## The parliamentary enquiry

I have become so out of touch with reality that I don't even notice that the clocks have gone forward. It is mid-March and still snowing. Moshannon Valley Correctional Center is situated on a plateau 3,000 feet above sea level. It is extremely cold here. So cold in fact that a mere pullover or sweatshirt is now a tradeable high-priced commodity. But now is not the time to falter. In a few minutes' time, I have a very important meeting in the visitation room.

I've been hoping for this moment for more than three years, and they are finally here, in front of me. Of course, I would have preferred to talk to them in different circumstances, I have so much to relate. Never mind, the main thing is they are here. Olivier Marleix (centre-right) and Natalia Pouzyreff (centrist liberal), Chairman and Vice-Chairman of the Parliamentary Enquiry Committee investigating Alstom, have travelled nearly 4,000 miles just to hear my story.

'Moreover,' Marleix tells me, 'the Americans didn't exactly make it easy for us; it took them more than a month to grant us a visitation right.'

Quickly, I notice that the two MPs are already very familiar with the background to the Alstom case. No need to convince them of American meddling in the affairs of large European multinationals. That would be tantamount to preaching to the converted. Two and a half years ago, Olivier Marleix actually attended the first conference on this theme at the Assemblée Nationale entitled 'After Alstom, who's next?' But their information was incomplete. For instance, they still do not know when the US investigation was launched, nor how Alstom was able to negotiate a timeframe in which to pay its fine. Over several hours, I answer all their questions, fill in the blanks, trace back the timeline and draw their attention to the disconcerting coincidences of dates.

They tell me about the interview they had the previous day in Washington with the Head of International Relations at the DOJ and his old acquaintance, Daniel Kahn, who has since been promoted in June 2016 from Acting Chief to Chief of the FCPA unit, following his successful handling of the Alstom case. Both were, of course, asked about their refusal to transfer me to France. The Head of International Relations preferred to dodge the issue, asserting that he was not familiar with my file. This cannot be true. I know that the French Ambassador and the Minister of Justice informed them directly of my detention. But what difference does one more lie make? More interestingly, the two MPs also questioned their interviewees about the leniency shown by the DOJ to Patrick Kron.

'Daniel Kahn replied that he did not have sufficient evidence to charge him,' said Marleix.

Another falsehood. All you have to do is consult Alstom's guilty plea to discover the contrary. The two MPs were determined to 'grill' my former CEO when they question him under oath in the Assemblée Nationale during their committee's proceedings. And they keep their word. From my cell, in the far reaches of Pennsylvania, without internet access, I am only able to follow this enquiry in a piecemeal fashion, via press cuttings sent to me by my family. The reports are very explicit.

A headline in the newspaper *Le Monde*, dated 5 April 2018, says: 'Takeover of Alstom by GE: Patrick Kron failed to win over MPs!' That's an understatement. In his opening remarks to the final report of the Parliamentary Enquiry Committee, Olivier Marleix takes an axe to the arguments raised by the former Chairman and

Chief Executive Officer. 'The line of defence adopted by Patrick Kron,' he writes, 'is one of evident deceit and untruths. Indeed, during his two previous hearings before the Economic Affairs Committee of the Assemblée Nationale on 11 March and 1 April 2015, he ruled out any association between the sale of the energy branch and the negotiations with the DOJ. But the truth is different. This is one of our Enquiry Committee's main findings.' Mr Marleix continues to labour the point: 'Did the threat of a monetary penalty weigh in on Mr Kron's decision to sell? Our Enquiry Committee answers this question in the affirmative.'

In the eyes of the Assemblée Nationale, Patrick Kron is therefore a liar. Does this worry him? Obviously not, because before the members of parliament, under oath, he still claims that he was never subjected to 'any pressure whatsoever, or any blackmail, neither from the Americans, nor from any other jurisdiction'. Then, when asked about my legal situation, he admitted (for the first time publicly) that I had 'done absolutely nothing in this affair for my own personal gain'. Given these circumstances, the MPs ask him why he fired me. And above all, why I received no compensation.

His reply was edifying. 'There was no opportunity to obtain a favourable outcome to this issue,' he replied coldly. Before daring to claim that he had 'done all he could do to help me'. According to him, all those who challenge the validity of his transaction with General Electric are merely spreading 'unfounded and insulting insinuations against him'.

However, before the Enquiry Committee, several witnesses contradict him. Arnaud Montebourg, the former Minister of the Economy, also testifying under oath, stated that he was convinced 'that physical pressure had been exerted on Mr Kron, such as the threat of his arrest'. A former Alstom senior executive agrees. Pierre Laporte, former general counsel of Alstom Grid, the group's branch specializing in electricity transmission, has a disturbing recollection to tell the members of parliament.

'In 2013, Mr Kron and Mr Carr met with the DOJ. Keith Carr, whom I saw the next day, then told me that he had phoned his two sons from the airport to warn them that he might not return from his next trip because the DOJ had threatened him and Kron with arrest.'

During the Parliamentary Enquiry he had initiated, Olivier Marleix also revealed an aspect of the Alstom/GE case that had been overlooked, i.e. the extravagance of the resources deployed by the two companies in media communication, financial arrangements and legal assistance. To complete the sale, Alstom used ten law firms, two financial advisers (Rothschild & Co.; Bank of America Merrill Lynch) and two communication agencies (DGM and Publicis). GE, for its part, used three financial advisers (banks Lazard, Crédit Suisse and Morgan Stanley), the communications agency Havas and numerous law firms. Alstom spent a whopping €262 million! It goes without saying that GE must have put a similar amount on the table.

'Does such an excess of funding still allow the government and shareholders to make informed decisions?' Marleix asks in his foreword to the Committee's final report. He pursues his rhetoric with vigour: 'Was there anyone left in the Paris marketplace who was willing to push through something that goes against those interests? Here are we not just witnessing remuneration for the mission undeniably carried out, but above and beyond, remuneration to influence the very decision itself?' I couldn't have expressed it better myself. I now understand why so few dissenting voices were heard at the time of the takeover. Silence is golden.

Finally, the Committee also points to the disturbing role played by Emmanuel Macron in this case. In October 2012, after being appointed Deputy Secretary General of the Élysée, fresh from Rothschild Bank (Alstom's financial adviser), he urgently requested a confidential study to be carried out. The request is written as follows: 'Evaluate the pros and cons of a change of shareholder in the company for French industry and employment.' This 'report,' writes Olivier Marleix, 'is based on accurate information relating to a change of majority shareholder'. Bouygues was Alstom's reference shareholder, with a 30 per cent stake, which it decided to sell. Therefore, concludes the MP, 'it is regrettable that the state authorities, who had enough accurate information at their disposal to commission a study at a cost of €299,000, at no time deemed it worthwhile to pursue their efforts in assisting Alstom, favouring a GE takeover scenario instead'. In other words, Marleix was convinced that Emmanuel Macron knew before anyone else what was being concocted.

In January 2019, Marleix reported to the prosecutor of the Republic of Paris the existence of a possible corruption pact concerning Emmanuel Macron. The file now in the hands of the Parquet National Financier (the French equivalent of the UK's Serious Fraud Office) alleges that some people who heavily funded Mr Macron's political campaign also benefited from Alstom and GE spending an extraordinary €600 million on consultants, financial advisers, lawyers and communication experts at the time of the takeover. I am obviously unable to know whether his analysis is relevant. Personally, I above all hope that Emmanuel Macron, who has since become President of the Republic, will write to Donald Trump to request a presidential pardon, as the information I receive on this subject is confusing. One minute, the liaison officer Navarri promises me that it has been carried out, and the next it appears to have been shelved. The timing is not good as Trump himself has big problems since his election in controlling the DOJ and the FBI. He even tweeted in August 2018 that 'an incredibly corrupt FBI & DOJ trying to steer the outcome of a Presidential Election', referring to the DOJ investigation over alleged Russian interference in the 2016 elections. He also stated in 2012 that the 'FCPA is a ridiculous and horrible law'!

Anyway, at the end of our meeting in the visitation room, Olivier Marleix promised to check on the situation with the French

Ambassador in Washington and Philippe Etienne, the President's diplomatic adviser.

Emmanuel Macron is due to arrive in the United States on 24 April. He will be the first foreign Head of State to meet Donald Trump on American soil since his election. The two presidents (who share very different political backgrounds) seem to like each other. Who knows, this could go in my favour. My imagination starts to run wild. What if Emmanuel Macron were to succeed in obtaining a presidential pardon? I could return to France with him? I'm getting carried away ... Wake up, Pierucci!

# 50

# Macron's visit to the United States

I shouldn't have got my hopes up. Emmanuel Macron did not write to Donald Trump about my situation. Though the French Presidential Palace did react. My family wrote to the President twice. My lawyer, Paul-Albert Iweins, former president of the Paris Bar, mobilized all his contacts, and many political figures gave me public support.

In the newspaper *L'Obs*, which Matthieu Aron joined after leaving Radio France, several former ministers came out of the woodwork.

'The American justice system has usurped inquisitorial rights against Frédéric Pierucci,' says Jean-Pierre Chevènement, previously Minister of Industry, Education, Defence and the Interior, under François Mitterrand and Jacques Chirac.

'I am in favour of his transfer to France,' he adds. 'We must release him; we have reached the limits of what is acceptable in this case and have even exceeded such limits.'

Pierre Lellouche points out: 'Pierucci is the perfect scapegoat. He did his job, and now he is paying for all those who set up these schemes.' The former Secretary of State for European Affairs and then for Foreign Trade in the Fillon government goes so far as to publicly worry about my safety. 'I'm afraid something will happen to him. Power is the only thing that the US Judiciary understands; it plays hard, it is out of control.'

Then Arnaud Montebourg adds: 'It is not Frédéric Pierucci who should be in prison, but his CEO, Patrick Kron, who is the real culprit in this whole fiasco.'

Daniel Fasquelle is even more severe.

'Alstom's management has lost interest in Frédéric Pierucci's fate. And Kron came out of it with a nice bonus. That's what shocks me the most: that Pierucci is in prison and Kron gets a cheque. The captain has abandoned his crew and ship.'

I risked media coverage of my case for the first time and it paid off. While Macron did not go so far as to ask for a presidential pardon, Justice Minister Nicole Belloubet, who accompanied him on his trip to Washington, did in fact speak with Jeff Sessions, the United States Attorney General. The liaison officer, Marie-Laurence Navarri, who was present at this meeting, defended my cause.

'How can you refuse the transfer of Frédéric Pierucci?' she pleaded with the American authorities. 'He satisfies all the required conditions: no homicide, no drug trafficking, no professional or personal ties in the United States, young children in France, no appeal of the sentence, $20,000 penalty paid, more than half of his sentence served in a maximum-security detention facility.'

She subsequently said that Jeff Sessions agreed and that I could reapply for a transfer. And that he had undertaken to examine it sympathetically. Which, in diplomatic language, means that this time they will respond favourably. A huge relief. This is just the beginning of a long and arduous struggle before I am released.

Navarri warns me: 'You'll have to wait for the DOJ's official approval, then arrange an appointment with an immigration judge [this can take several weeks], then you will be transferred to a prison in Brooklyn, or Manhattan [which again can take several weeks]. Finally, you will be sent back to France.'

And once I arrive in Paris, it's not over. As soon as I disembark from the plane, I will be presented to a prosecutor, then placed in detention in a French jail, before requesting my release on parole. Yes, all this will take months. But even if I only gain one extra day of freedom, it will be worth it, as this place is dire.

I had hoped with the arrival of spring that tensions would ease. But not so. Just yesterday, one of our own, an 'International', a Georgian national, came close to being lynched by Mexicans, who accused him of lack of hygiene. Even Muay Thaï, the ex-legion-naire, is not out of their reach. Given that he wakes up at 3 a.m. to watch the UFC matches (a mixture of martial arts and combat sports) quietly, almost alone, the Mexicans blocked that channel, even though there was nothing else on at that hour.

Everything seems to have become darker, more gruesome, more violent, unless my resistance is weakening. Sometimes I freak myself out. Every night I have a recurring nightmare in which I am in a very long tunnel, searching for the end which I can never see . . . And I live in fear of being sent to solitary in the 'hole'.

There are no secrets in prison, and the rumour is already starting to circulate that I could be transferred soon, which is bound to create jealousy. There are many stories going around. I even hear that envious inmates pay sometimes for the services of a penniless prisoner to beat up the one who is leaving. All involved in such machinations are put in the 'hole' and an investigation (which usually lasts three months) is opened. This delays the transfer even further. Many therefore hide their release date to avoid attacks.

In addition, the MVCC administration are giving me a hard time. Perhaps they are annoyed at the French government's inter-vention? It's like they are making me pay for receiving assistance.

For two weeks now at least, they have refused to pass on the newspapers sent to me by my family and friends. I received four notices from the Department of Justice/Bureau of Prisons indicat-ing that the 'content of my letters was not authorized'. Usually, we receive this kind of warning when the envelope contains sexually explicit photos or stamps with drugs attached behind them.

I went to their office and was greeted by a moron, one of those guards who gets his kicks out of being sadistic to prisoners for trivial matters. He asked me to make a choice: either he would destroy the newspapers, or he would return them to the sender at my expense. No matter how hard I tried to make him listen to reason, the tone quickly rose. But it wasn't over. This idiot then showed me about ten holiday photos with Gabriella and Raphaella that my friend Leila had just sent me. According to him, he can't give them to me because they're not in the American 5 × 7 format, the only size allowed for photos. I tried desperately to explain to him that in Europe the format was different, but he didn't want to hear. Faced with so much nonsense and bad faith, I started to get very upset. Fortunately, another guard intervened, just before I was about to have disciplinary action taken against me.

Though cautious about the rules imposed on inmates, the GEO group, owner of MVCC, should perhaps exercise better control over its staff. One member of the management team had to resign overnight. There was a rumour in the prison that bribes were paid by suppliers (which shows that it doesn't only happen at Alstom . . .). I don't know if the rumour is true, but I admit I find it satisfying to hear. Another far more dramatic piece of news that came to our attention was that in a prison very similar to ours in South Carolina in mid-April, fights between prisoners broke out and left seven dead and seventeen seriously injured. The prison staff let it play out for seven hours without intervening. This did not move the Governor of the State, Henry McMaster, much at all: 'We know that prisons are places where people who have not behaved well are placed. It is therefore not surprising to see them behaving violently,' he told the *Washington Post*. How cynical can you get? Is a man's life, however miserable, worth so little? And to think that these comments are made in the country with the highest incarceration rate in the world. More than China, India and Saudi Arabia. In 2012, 2.2 million people were imprisoned in the United States. That's 25 per cent of all prisoners in the world. This number makes me dizzy. In France we incarcerate ten times fewer. One in three blacks in the United States will go to prison at least once in their lives.

And in Moshannon Valley and Wyatt, many prisoners have difficulty reading or writing. I therefore helped them to write their administrative requests, or even to draw up business plans for their future. Some prisoners who have successfully purchased land in their countries of origin want to engage in legal activities upon their return, once deported from the United States. Such as the Mexican who wants to export mangoes to Canada, the Dominican who is looking for outlets for his cocoa plantations, or the Ghanaian who created an organic market before being arrested.

I meet up regularly with this small group of budding entrepreneurs in the library. It keeps me busy and makes me feel somewhat useful. Even though I have trouble creating my own business plan. Summer is coming and I am still waiting for the DOJ to formalize the promise it made to Justice Minister Nicole Belloubet to authorize my transfer. I'm still in a state of uncertainty. Besides, I lost my lawyer. Stan Twardy has made it clear that I can no longer count on him. As I have had my final sentencing, and given that I can no longer pay his bills, Stan considers that he is no longer under any obligation to defend me. It's debatable, but given my perception of his defence, I don't consider him a loss.

# 51

## The arduous journey to freedom

In Moshannon Valley, out of 1,800 inmates, there is only one Frenchman: Frédéric Pierucci. Therefore, on 13 July 2018, day of the football World Cup Final, I am the hero of the day. For once, there were no fights over the TV programmes. The whole pod is glued to the screen. I now have the support of the Africans, the Russians, the Canadians and the Romanians. The Mexicans, on the other hand, are cheering on Croatia. An explosion of joy at Pogba and Mbappé's goals, a little fright after Lloris's blunder, and then the great feeling of pride when France win. There is a good atmosphere and it reminds me of 1998. At 3 a.m. I had watched our previous national victory with other French expats in Beijing, where I was stationed. I am a little more relaxed than I have been, since I learned at the beginning of the month that the DOJ had officially authorized me to return to France. Relaxed, but on my guard. I live in permanent fear of them finding a way to force me to stay by implicating me in a new case. Who knows, they might force me to stay another year on American soil on 'probation'. This has never been done in a case comparable to mine, but they are capable of any kind of turpitude.

What I fear most are the 'snitches', the prisoners who work under cover for the FBI and abound in Moshannon Valley. I've had two alerts in the last two weeks. First, a Georgian who was arrested in a major narcotics case in New York, who has just joined our pod. Two of my fellow inmates caught him going through my papers. There are a lot of Russians in C5 at the moment and they carried out their own investigations via their external contacts. They got confirmation that the Georgian was indeed a snitch. The representative of the 'Internationals' informed the management, who immediately removed him to another block.

And then, a week later, we found another snitcher who was also snooping around me. This time the Moshannon administration sent him directly to the hole to protect him. And that's not all. I also recently received a very strange letter from a prisoner in Wyatt, whom I had met during my 'stay' there. My former cell mate obviously knew that it was strictly forbidden for prisoners to write to each other and that I could be punished for receiving his letter. Why did he put me in this situation? What is his motive? Is he also a snitch? Stop seeing conspiracies and schemes everywhere. Don't get paranoid. What if I already am paranoid? I have to get out of here quickly, otherwise I'll go insane.

On 25 July I will be speaking to an immigration judge via a video link to confirm my request to be transferred. Then I only have to wait for my actual departure in three to six weeks. However, at the last moment I am gripped by doubt. Given the snail-like pace of the American administration (which I am convinced is deliberate), is it not in my interest to wait until the end of my sentence? At the end of October, beginning of November, I will be released and this would avoid me finishing my sentence in France and thus getting a French criminal record. Suddenly, reason takes over. No, just get out. Get out of here as quickly as possible!

It's 9 September 2018. This afternoon, I anxiously checked to see if my name was on the exit list posted in the pod's corridors. It is indeed. Huge sigh of relief. I had to be patient. The American penitentiary exhausted all possible means to ensure that I was transferred at the last possible moment. But everything has an end, even the worst nightmares. I am scheduled to leave tomorrow at 8 a.m.

At dawn, the guards ask us to strip. Former legionnaire Muay Thaï will also be travelling. He is being deported back to Slovakia.

We must dress in the 'transfer attire' reserved for inmates in transit, i.e. T-shirt, khaki trousers, canvas sandals. Then, with five other prisoners, we board a bus in the pouring rain, after having our hands and feet shackled. Fortunately, they didn't put the steel bar between our wrists that cuts your skin and digs into your joints, as the journey to New York is going to be a long one. It should take about eight hours. The air conditioning is going at full blast in the bus and we shiver in the arctic temperature. Despite our repeated requests, our guards, warmly wrapped in their parkas, stubbornly refuse to raise the temperature. Just before midday, our convoy stops in the Harrisburg airport cargo area usually used by the army.

At the edge of the track, about fifteen buses similar to ours, and a multitude of SUVs and small vans are waiting for planes to arrive. I discover that, once a week, Harrisburg is transformed into a sorting station, through which all inmates who are transferred between the various American prisons transit.

When an aircraft taxies to a halt, dozens of heavily armed police officers wearing bullet-proof vests, machine guns at the ready, take their positions around the walkways.

As the rain continues to fall in torrents and the night begins to fall, I am forced to walk faster in chains and canvas sandals, to the cries and shouts of the guards, on the very slippery runway with the sensation of being in a horror movie. Like a cursed wretch, I take small steps towards the planes; then, at the last moment, a guard pulls me out of the line, and pushes me towards a bus. Contrary to what I thought, I am not boarding a plane bound for New York. Once I'm in, the vehicle sets off. One of my fellow inmates (who has already experienced this type of transfer) informs me of our new destination: the Canaan High Security Penitentiary in north-eastern Pennsylvania.

We arrive early evening. The admissions process is tedious and interminable and takes almost four hours. Finally, we collapse into our cells, famished and dehydrated. We haven't drunk anything since this morning. Upon waking up, we learn that we will have to spend twenty-four hours in this place before leaving for Manhattan. I only have one memory of Canaan: the vile food – impossible to ingest anything in that place. It must be said that we inmates are wary of each other. A few years ago, in 2011, in this particular prison, more than three hundred prisoners, and also guards, fell ill with the most serious salmonella poisoning in the history of the United States.

At about 10 p.m. we hit the road again, this time heading for the Metropolitan Correctional Center in Brooklyn. A new stopover between 1 a.m. and 5 a.m., during which we remain locked in a cage, like cattle, with thirty-six other detainees, namely four Hispanics, and thirty-two black Americans. Muay Thaï and I are the only two whites. Then this haunting journey ends. It lasted more than three days. Three days to cover less than 250 miles.

On 12 September 2018, still handcuffed and shackled, I entered the Metropolitan Correctional Center in southern Manhattan. I am overcome by trauma, as it was in this very same correctional center that I spent my first horrifying night, right after my arrest, and my first interrogation at FBI headquarters on 14 April 2013. That was five and a half years ago.

Metropolitan Correctional Center, like Wyatt, is a maximum-security prison. The US press have dubbed it the 'Guantanamo of New York'. It is here that the most heinous criminals are detained, awaiting trial or extradition to their countries of origin. In the quarter where I am assigned, the cell opposite mine is occupied by a triple murderer, the one on my left by a Bengali who was arrested a few months ago wearing a suicide vest. He tried to blow up the New York subway. And in the cells just below, there are two lieutenants of the Mexican drug lord El Chapo, his main hit man, a *sicario* accused of killing 158 people, and his 'banker' who was in charge of laundering the narco-trafficking proceeds. El Chapo, himself, is held a few floors above me, in solitary confinement.

I don't have time to settle in as I am called to the visitation room. A pleasant surprise awaits me there. Jérôme Henry, the Deputy Consul of France in New York – accompanied by Mrs Hélène Ringot, the head of social services – wanted to see me as soon as I arrived at the Metropolitan Correctional Center. After two all-nighters and with no possibility of bathing, I am in a pitiful state, but so be it. I am delighted to see them. We discuss all the practical details of my transfer to France. The immigration agents have allegedly 'lost' my passport, so will have to provide me with a special pass. Strange as it may sound, it is Jérôme Henry who will bring me the clothes that Clara purchased online and had delivered to the Consulate. All I have is the dirty T-shirt on my back, which I have worn for the last three days, and a pair of broken canvas sandals, which force me to walk half barefoot.

I am going to have to be patient another eight days at the Metropolitan Correctional Center, waiting for my return to France, which is not scheduled until 21 September. I have to survive a whole week in this cesspool of grisly murderers and would-be terrorists. The sanitary conditions are terrible. Dampness infiltrates everywhere. All the water pipes leak, while most of the showers are out of service and have been for a very long time. As for the WCs, they are permanently blocked. On our floor, there was a cell whose door no longer closed, so it was abandoned and turned into a waste dump. The stench emanating from it was foul. But the worst part is at night, when the place is crawling with mice. They are aggressive and bite your face in your sleep. For this reason, everyone covers their head with a blanket. To add to this, I am penniless. The money I had in my commissary account in Moshannon Valley has not been transferred to the Metropolitan Correctional Center. I cannot purchase anything: no bowl, glass, spoon or sandals. We are all in the same predicament and have to manage with the means at our disposal. Finally, we find a single pair of sandals, which we share among us, taking turns to use them.

The days drag by very slowly. To kill time, I do maths exercises and I help a young Haitian man who hopes one day to pass his secondary education diploma. Finally, 21 September arrives. French prison administration officials collect me directly from the Metropolitan Correctional Center before escorting me, together with the US Marshals, to JFK airport to board an aircraft bound for Paris. However, right up to the last minute, I fear that my transfer will be cancelled, since that same day, El Chapo is escorted to the courts for his trial. Hundreds of American police officers had blocked the entire Metropolitan Correctional Center neighbourhood and the Brooklyn Bridge with a spectacular show of force.

Abruptly, three hours before the departure of my flight, I was extracted from my cell, shackled from head to toe and thrown into an armoured vehicle. Our convoy then sped down the streets of New York, sirens wailing, to reach the airport in time. It was only at the foot of the walkway that I was officially handed over to the French authorities.

I am now safely on board the Air France plane. Three prison officials have come to escort me. My case was reported to them in Paris; they know that I am not dangerous and so they quickly remove my handcuffs. We are seated in the last row, we talk, and I almost feel like a free man.

At 5.30 a.m. we land at Paris's Charles de Gaulle airport. I want to kiss the ground. When I disembark from the plane, I am taken to the Bobigny courtroom and, in compliance with the procedure for transferred prisoners, I am immediately referred to the public prosecutor. I am placed in a cell until a *juge d'application des peines* (judge for sentence enforcement) examines my case.

At that point, I even hoped to be released later that day. But, unfortunately, there were no judges on duty. After twenty hours of waiting in the cell, I was taken to the Villepinte prison, for the weekend, in the hope that a judge would examine my case the following Monday. I am given a warm and professional welcome. To ensure my safety, I was offered my own cell. I leapt at the chance, as you can imagine after one year spent in a dormitory. At last some privacy. Finally, some comfort. The cell is spacious, there is a TV, separate toilets, meals are very good and the guards are particularly courteous. They are clearly giving me 'preferential' treatment in the true sense of the term. Moreover, I learned that Olivier Marleix, the Chairman of the Parliamentary Enquiry Committee who had launched an investigation into Alstom, had come to the Villepinte prison the evening I arrived, hoping to meet me. But I was still locked up in the Bobigny Court of First Instance at that time. Then, Monday morning, just seventy-two hours after touching down on French soil (i.e. within a very short judicial period), the judge responsible for sentence enforcement examines my case. He grants me immediate parole.

On Tuesday 25 September 2018, at 6:00 p.m., five and a half years after my arrest at JFK Airport, after spending twenty-five months in United States jails, including fifteen months in a maximum-security detention facility, I am released from prison. Freedom at last ...

# Epilogue

At the time I'm finishing writing this book, with Matthieu Aron, I have been free for five weeks. My family, my partners, my friends, all advise me to take a break, go away, relax.

It's not the right time. I don't want to be like those prisoners who, broken by the trauma of their prison experience, try to replenish their strength, or, feeling lost faced by the blank page that opens in front of them, attempt to forget all by forging themselves a new path. I don't want to 'move on'. I want to continue the fight. I want to serve. This is a war.

François Mitterrand, at the end of his term of office, confiding in Georges-Marc Benamou, was right when he said: 'France does not know it, but we are at war with America. Yes, a permanent war, a vital war, an economic war, a war without death, on the surface. And yet a war to death.'

This war is not *my war*. It is *our war*. This is a war that is more sophisticated than conventional warfare, more insidious than industrial warfare, a war that the public are unaware of, a war of the *law*. Experts at the Centre for the Analysis of Terrorism have described this new type of warfare, known as 'law fare', which consists of using the legal system (the law) against an enemy, or an adversary designated as such, in order to delegitimize such an adversary, causing it maximum damage and forcing it to comply using coercion.

This concept was rendered official shortly after the attacks of 11 September 2001, by US army colonel Charles Dunlap. It has since been echoed by many researchers in American neo-conservative circles, who have advocated broadening its area of application. Indeed, the United States has succeeded in imposing a set of standards on its allies and companies on consensual subjects such as the fight against terrorism, the fight against the spread of nuclear weapons and the fight against corruption and money laundering.

All these are legitimate and necessary combats, but they have allowed the Americans to proclaim themselves 'global police'. By virtue of the power of their dollar (used for worldwide trade) and their technology (which allows data to be transmitted globally via their internet messaging systems), are they really the only ones in a position to enact extraterritorial laws and, above all, enforce them? From the late 1990s, European countries agreed to submit to *lex americana* (US law). Yet, to date, they have been unable to devise similar mechanisms to defend themselves or to retaliate. Do they even have the will to do so?

For almost twenty years, Europe has allowed itself to be held to ransom. The largest corporations in Germany, France, Italy, Sweden, the Netherlands, Belgium, England and Sweden have been convicted one after the other for corruption or banking offences, or for violation of embargoes. And the US Treasury has netted tens of billions of dollars from this. It has taken more than $13 billion from French companies alone. Not to mention all those that will inevitably fall into the clutches of the US justice system tomorrow. Starting with our two ultra-strategic multinationals, Airbus and Areva (renamed 'Orano'), both of which are also in the DOJ's firing line for corruption cases.

This racketeering, because that is what it all boils down to, is unparalleled in its scope.

At the beginning of 2019, I am having trouble containing my anger in light of what happened to Alstom and its employees. Not one of the promises made by Jeff Immelt, the CEO of General Electric at the time of the takeover, has been honoured. Not one.

The pseudo joint ventures, so touted by our government, have turned out to be a mere pipe dream. Furthermore, GE will not deliver on its pledge to create a thousand new jobs in France. The group has already announced the slashing of 354 jobs out of 800 in Grenoble, while in Belfort, subcontractors are noticing that the orders promised have not materialized. From 2019 onwards, employees of Alstom, as it was, will no longer be protected from the massive job cuts to be made by GE in Europe (4,500 jobs cuts announced, i.e. 18 per cent of its workforce). And this is just the tip of the iceberg. In fact, GE ended up announcing on 28 May 2019, just two days after the European parliamentary elections in France, that they would cut more than 1,000 jobs, mainly in Belfort, instead of creating the 1,000 promised.

On 30 October 2018, Larry Culp, GE's newly appointed CEO, announced a write down of \$22 billion in the third quarter, and a restructuring of the energy division. It seems like a long time ago that Patrick Kron was bragging on TV and radio stations about the 'fantastic industrial venture', promising that jobs would be 'safe-guarded' and a bright future with 'major advances in the energy sector'. Some commentators are even trying to rewrite history. According to them, the setbacks experienced by GE are an argument in favour of the sale of Alstom. Patrick Kron, according to them, is an incredible visionary. He supposedly anticipated what was going to happen, before anyone else, and even 'pulled the wool over GE's eyes'!

Just who do they think they are kidding? First, it is usual for a new chief recruited to redress a company in difficulty, to strike hard and fast upon his arrival, announcing significant losses due to the 'misguided choices of his predecessors', in order to be able to quickly show signs of improvement. Thereafter, any power sector specialist understands that the evolution of the energy market is cyclical in nature, experiencing overall growth over the long term. These so-called 'commentators' give a somewhat simplistic analysis of the difficulties encountered by GE. No, GE's problems do not date from its acquisition of Alstom, not by a long shot.

Its share price has fallen by more than 75 per cent since September 2000. GE was on the verge of bankruptcy following the sub-prime crisis of 2008, which hit its financial subsidiary, GE Capital, head-on. Since then, GE had accumulated vast debt, which it has not been able to discharge completely, mainly from this moribund division.

And then, when we look carefully at the announced loss of $22 billion, we see that it is in fact a pure accounting phenomenon, resulting solely from the goodwill impairment of GE's entire Power division with no impact on cash. And not just from the acquisition of Alstom Power in 2014 but since the creation of the GE Power business. This strange accounting move even drew the attention of the DOJ and of the SEC who both extended an ongoing investigation into GE accounting practices originally targeting its legacy insurance liabilities.

That said, it should be noted that GE Power had an extremely well-filled order book worth some $99 billion, representing more than two and a half years of work. But since then the gas turbine market where GE had held historically the number-one position collapsed whereas the nuclear market where Alstom was the leader has boomed. The situation of the GE Power business originating from Alstom is thus not that catastrophic at all, unlike that of the original perimeter of the GE Power group. Another explanation can be sought elsewhere, particularly in terms of technological issues. In September 2018, GE announced oxidation problems potentially affecting fifty-five of their new gas turbine model, already delivered to its customers.

On the other hand, this is the end for Alstom. In 2019, its transport business was in the process of being acquired by Siemens. The European Union rejected this takeover in February 2019. Alstom Transport still exists but for how long? But the economic destabilization of this company, which I served for twenty-two years, is not an isolated case. Let us look at the situation in Iran. How can we stand by and watch our largest industrial groups pull the plug on the huge new contracts they have secured in that country, simply because Donald Trump suddenly decided to withdraw from the nuclear agreement, and unilaterally reimpose an economic embargo on Tehran?

Total, which was supposed to operate 50 per cent of the world's largest gas field, and Peugeot SA, which had planned to build 200,000 cars per year, have already had to withdraw. They were afraid of being prosecuted by the US judiciary if they continued to trade with the Iranians. In the name of what overarching principle can an arrogant billionaire president impose such a diktat?

I can see that some nations are trying to react. For instance, Germany's Minister of Foreign Affairs, Heiko Maas, is urging its European partners to adopt a non-dollar payment system to avoid FBI prosecutions. As is Bruno Lemaire, the French Minister of the Economy, who rebels against arbitrary 'Trumpian' edicts. 'Do we really want to be vassals of the United States, blindly obeying with a curtsy and a bow?' he declared in May 2018 on the subject of Iran. It is high time we turn our words into deeds.

Especially since the threat is increasing. The United States has enacted in March 2018 the 'Cloud Act' which virtually 'legalized' the economic espionage of the NSA Prism project. This law allows federal law enforcement like the DOJ 'to compel U.S.-based technology companies (like Microsoft, Amazon or Google) via warrant or subpoena to provide requested data stored on servers regardless of whether the data are stored in the U.S. or on foreign soil'. Hence the DOJ can bypass international judicial cooperation with foreign ministries of justice and access directly information on any company using the cloud services of these US tech giants. It therefore enables US intelligence agencies to gain easier access to personal and company data stored outside the United States. Emails, online conversations, photos, videos, confidential corporate documents. All such information, depending on their political or economic strategies, can be used to 'feed' the American intelligence services' database. Yes, our leaders must now display political courage to avoid us becoming – in Minister Lemaire's words – permanently reduced to 'vassal' status. Imagine what the American reaction would have been if France or another European country had incarcerated an American Google executive for tax fraud? Should it come to that to gain respect? If we continue to remain passive, other countries, such as China, may also soon impose their own extraterritorial laws on us.

Prompt action must therefore be taken, at European level. For example, as proposed by former Prime Minister Bernard Cazeneuve, now a lawyer, by implementing a European anti-corruption public prosecution service. This is the only legal instrument that would be powerful enough to compete on equal terms with the DOJ.

Though let us not delude ourselves. President Obama's administration has been the most active in using the law as an economic weapon against European companies while at the same time calling an unprecedented war against domestic whistle-blowers like Edward Snowden or Chelsea Manning who, with heroic courage, exposed to the world the lies and the illegal actions of the US government. But whoever is the US president in power, Democrat or Republican, charismatic or obnoxious, the Washington administration always serves the interests of a plutocracy who own its industrial, financial and more recently technological giants. To put it in a nutshell, the personality of the individual sitting in the Oval office determines whether US extraterritoriality is palatable to us or not. And we forget, or we close our eyes to the fact, that the United States, which takes the moral high ground, telling the whole planet how to behave, is the first to conclude phony deals in many countries under its area of influence, starting with Saudi Arabia or Iraq.

But right now, the context is somewhat different. The ground is fertile for developing a sense of awareness, even consciousness. Today, the brutality of American unilateralism and imperialism is staring us in the face. Let us not miss this golden opportunity.

It is now time for Europe and France to redefine our foreign policies and stand up to the United States to regain our economic, political and judicial sovereignty.

# Afterword

By *Alain Juillet, former Intelligence Services Director of the DGSE (French Counter-intelligence Services), former Senior Official for Business Intelligence, President of the Academy of Economic Intelligence*

After the BNP Paribas and Total cases, Alstom's dealings with the American judiciary raised many criticisms and questions. Enquiry committees at the Assemblée Nationale and the Senate were set up by elected officials to try to understand how France could have let one of its industrial icons slip away. Above and beyond the pacifying statements made by Alstom's CEO, who repeatedly denounced a conspiracy campaign against him, there were indeed missing pieces to the puzzle, since the senior officers of Alstom and General Electric had been careful not to disclose everything to their boards of directors and elected officials. The sad reality is that the company breached its obligations and – as we learn from this book – continued to do so despite numerous warnings.

On reading this book, we get a better understanding of the reluctance of the company's top management to admit to something so shameful. Aware that they risked prosecution on grounds of corruption and aiding and abetting the corruption of foreign public officials, some sought to save their own skin by sacrificing others.

The fact of the matter is that, for the last decade, European enterprises have been in the cross hairs of the United States Department of Justice. Not only are they heavily fined, but they are also placed 'under supervision'. Americans are not just content to cash in mega-bucks from these firms; they also impose the presence of a 'monitor' on these companies over several years.

These so-called 'monitors', which are appointed by the DOJ but paid for by the company being sanctioned, are supposed to verify that compliance procedures are being respected. Except that these standards, set according to criteria specifically designed by the United States authorities, do not necessarily correspond to our vision of professional ethics. Nor to our vision of ethics at all. Let us hope that with the arrival of the Sapin II law, which should enable us to tackle corruption more effectively while protecting our businesses, we will see some positive changes.

After reading this book, our public and private corporate leaders have the information at their disposal to truly understand the methods and practices deployed by the United States to exert their economic dominance and achieve their goals. By means of a series of consecutive laws, our American friends have gradually broadened the scope and interpretation of the fight against corruption. With the aid of their intelligence services, they have set up a 'war machine' that enables them to prosecute all those who do not comply with their unilaterally made rules. Admittedly, it is easier to be 'global policemen' when you have the full backing of the NSA and the extent of its surveillance methods.

Admittedly, ignorance of the law is no excuse. But the extraterritorial nature of US anti-corruption law remains highly contentious. All the more so, since this extraterritoriality is not reciprocated. As a consequence, many international lawyers consider it abusive and forced. And what is true of the FCPA mechanism is also true in other domains. The United States does not hesitate equally to punish all those who prefer to buy weapons from their Russian or Chinese rivals, or those who want to trade with a country that the US has embargoed.

Confronted with this imperial logic that relies on military prowess, the law and digital capacity, the opponent has no choice: he must either submit, cooperate or perish. In the face of these practices, we must be realistic and stop daydreaming. As Churchill so rightly said, we have no friends. We only have enemies, competitors and partners. Far from the *hard power* of President Bush junior, the *smart power* of President Clinton, and the *soft power* of President Obama, we are now in the midst of *tough power*, and this is only the beginning. Is it right that our governments and those of other European nations do not have the means to counter-attack? Have we become so weak that all that remains within our reach is an undignified withdrawal?

What Frédéric Pierucci has experienced and in turn recounted with great flair is more than just a first-hand testimony; it is a genuine portrayal of the twenty-first century. Although his personal nightmare is now over, there are still other French companies out there who are ignorant of the cruel reality of international competition and the practices of certain countries, and thus remain vulnerable and exposed. Let us hope that this book will open their eyes and make them think. Frédéric's ordeal will then have served some purpose.

# Appendix 1: Monetary penalties inflicted on European banks by the United States

Over the last ten years, fines inflicted for non-compliance with international economic sanctions enforced by the United States have essentially hit European banks.

The only American bank to be punished, albeit with leniency, for embargo violation appears to be J. P. Morgan Chase. Since 2009, European banks have shelled out around $16 billion in fines to US authorities.

French bank, Société Générale, should be added to this list, as in June 2018 it had to pay more than $1 billion to the DOJ and the Commodity Futures Trading Commission (CFTC) to settle two litigation cases pertaining to fraudulent use of the Libor interbank rate and allegations of bribery in Libya. And in November 2018, it had to pay $1.3 billion to the DOJ and the US General Reserve for violating embargoes against Cuba.

The highest fines inflicted for violation of US international sanctions and/or money-laundering laws* are:

---

\* Note taken from the information report of the Foreign Affairs Committee and the Finance Committee of the Assemblée Nationale on the extraterritoriality of US laws, dated 5 October 2016.

| Company | Country (of registered office of parent company at time of the offence) | Aggregate amount (FOCA, DOJ and/or Fed and/or State and County of New York) of fines paid in the United States ($m) | Year of transaction |
|---|---|---|---|
| BNP Paribas | France | 8,974 | 2014 |
| HSBC | United Kingdom | 1,931 | 2012 |
| Commerzbank | Germany | 1,452 | 2015 |
| Crédit Agricole | France | 787 | 2015 |
| Standard Chartered | United Kingdom | 667 | 2012 |
| ING | Netherlands | 619 | 2012 |
| Crédit Suisse | Switzerland | 536 | 2009 |
| ABN Amro/ Royal Bank of Scotland | Netherlands | 500 | 2010 |
| Lloyds | United Kingdom | 350 | 2009 |
| Barclays | United Kingdom | 298 | 2010 |
| Deutsche Bank | Germany | 258 | 2015 |
| Schlumberger | France/United States/ Netherlands | 233 | 2015 |
| Clearstream | Luxembourg | 152 | 2014 |
| UBS | Switzerland | 100 | 2004 |
| JP Morgan Chase | United States | 88 | 2011 |

# Appendix 2: How General Electric hushes up its corruption scandals

In 2008, Andrea Koeck, a lawyer in the Consumer and Industrial division at General Electric, alerted her managers. She had uncovered an in-house VAT fraud mechanism and claimed to have also exposed some unscrupulous practices (bribes) linked to contracts awarded in Brazil.

So, what do those higher up the ranks do? They fire the messenger who brought the bad news. Then, when the press got hold of the scoop, General Electric, which professes to be 'a champion campaigner in the fight against corruption', got out its cheque book to silence the whistle-blowing lawyer.

Another similar case is the Asadi affair, named after the Chairman of General Electric in Iraq. In the summer of 2010, Khaled Asadi objected to the hiring of Imam Mahmoud, a woman with very close ties to the Iraqi Deputy Minister of Electricity. Asadi refused to give her a 'dummy function' in exchange for General Electric having obtained a $250 million contract. Not long afterwards, Asadi reported this to his managers. Like Andrea Koeck, he was pushed out and then forced to resign.

The former chairman of General Electric in Iraq then filed a suit and sought application of the Dodd–Frank Act, which protects whistle-blowers in the United States. The US judiciary dismissed his claim, however, on grounds of the alleged facts having occurred abroad, which is out of the scope of the Dodd–Frank Act. The United States therefore considers that they enjoy international jurisdiction when it comes to prosecuting companies, but not when it comes to protecting whistle-blowers.

# Appendix 3: A study of fines paid to US authorities under the FCPA (>$100m)

| Company | Country | Date | Sanction US DOJ+SEC (millions) |
|---|---|---|---|
| MTS | Russia | 2019 | $850 |
| SIEMENS | Germany | 2008 | $800 |
| ALSTOM | France | 2014 | $772 |
| TELIA | Sweden | 2017 | $691.6 |
| KBR/HALLIBURTON | USA | 2009 | $579 |
| TEVA PHARMACEUTICAL | Israel | 2016 | $519 |
| OCH–ZIFF CAPITAL MNGT | USA | 2016 | $412 |
| BAE | UK | 2010 | $400 |
| TOTAL | France | 2013 | $398.2 |
| VIMPELCOM | Netherlands | 2016 | $397.5 |
| ALCOA | USA | 2014 | $384 |
| ENI/SNAMPROGETTI | Italy | 2010 | $365 |
| TECHNIP | France | 2010 | $338 |
| SOCIÉTÉ GÉNÉRALE | France | 2018 | $293 |
| WALMART | USA | 2019 | $282.7 |
| PANASONIC | Japan | 2018 | $280 |
| JP MORGAN CHASE | USA | 2016 | $264 |
| ODEBRECHT/BRASKEM | Brazil | 2017 | $260 |
| SBM OFFSHORE | Netherlands | 2017 | $238 |
| FRESENIUS MEDICAL | Germany | 2019 | $231 |
| JGC Corporation | Japan | 2011 | $218.8 |
| EMBRAER | Brazil | 2016 | $205.5 |
| DAIMLER | Germany | 2010 | $185 |
| PETROBRAS | Brazil | 2018 | $170.6 |
| ROLLS-ROYCE | UK | 2017 | $170 |
| WEATHERFORD | Switzerland | 2013 | $152.6 |
| ALCATEL | France | 2010 | $138 |
| AVON PRODUCTS | USA | 2014 | $135 |
| HEWLETT PACKARD | USA | 2014 | $108 |
| KEPPEL OFFSHORE & MARINE | Singapore. | 2017 | $105 |

| | in million $ |
|---|---|
| EUROPE | $6,419.9 |
| US | $2,164.7 |
| OTHERS | $1,758.9 . |

Source: IKARIAN
(As of 30th June 2019)

# Appendix 4: The difference in treatment of Dow Jones 30 and CAC 40 companies for FCPA violations

## DOW JONES 30

Department of Justice:
Three companies
- Johnson & Johnson 2011
- Pfizer 2012
- JP Morgan 2016

Securities and Exchange
  Commission:
Two companies
- IBM 2000 and 2011
- Dow Chemical 2007
No indicted employees
PENALTIES: $343 million

(As of 30th April 2019)

## CAC 40

Department of Justice:
Five companies
- Technip 2010
- Alcatel 2010
- Total 2013
- Alstom 2014
- Société Générale 2018

Securities and Exchange
  Commission:
One company
- Sanofi 2018
Six indicted employees
PENALTIES: $1,965 million

# Acknowledgments

To my mother and Roland, my father and Anne-Marie, my sister and my brother-in-law, who all put their personal lives on hold for five long years to help and support me, my wife and our children.

Special thanks to Linda and Paul, and Michael and Shalla, without whom I would never have been released in June 2014. They put up their homes as bail bond. I will be indebted to them for ever. I wish to thank them here for their generosity and their faith in me.

Thank you to all my friends who supported me throughout this ordeal:

Tom for his loyal and unwavering friendship and his warm hospitality in 2014; Antoine and Claire, Leila and Stany, Didier and Alexandra for their loyalty in the face of adversity, and being there for my family at all times; Paul-Albert for his mobilization and his talent; Markus for his professionalism, his commitment to my cause and his self-sacrifice; Pierre for his availability, his genuine concern and for 'walking' to Moshannon to visit me; Leslie, Eric, Loik and Claude for their pioneering efforts in bringing this scandal to the attention of the public and for their moral support; Deniz for her ongoing optimism and kind patience towards me; Jean-Michel for his constant concern for my situation, his letters and the connection he maintained with former Alstom colleagues; Philippe the only member of the Alstom Executive Committee who didn't turn his back on me; François and Amy for their unconditional support at a time when very few people rallied around; Gilles and the whole Taylor Wessing team for having believed in me and for 'accommodating me' at their offices for two years.

Thanks to Laurent Laffont for the trust he placed in me, and to Paul Perles for his meticulous proof reading.

Special thanks to Olivier Marleix for his in-depth knowledge of the case and his brilliant and steadfast chairmanship of the Parliamentary Enquiry Committee; and to Natalia Pouzyreff, vice-chairman of this Committee. They both undertook a long journey to interview me at Moshannon.

Thank you to ministers Arnaud Montebourg, Pierre Lellouche, Jean-Pierre Chevènement, and MPs Daniel Fasquelle and Jacques Myard for their support.

Thank you to all those who had the opportunity to hear about my case in the course of their work and whose support has been invaluable to me, notably Marie-Laurence Navarri for her tenacity, her relentless commitment and her efficiency.

Thanks also to Céline Tripiana for her commitment to my case; to Jérôme Henry for his long-term support, his efficiency and his immediate understanding of my situation; to Hélène Ringot, Simon Cicollela, all of whom are a credit to the consular service; to the interministerial delegate for economic intelligence, Claude Revel; and to members of parliament Marguerite Deprez-Audebert and Stéphanie Kerbach.

Thank you to all those who wrote to me and visited me at Wyatt and Moshannon or who provided support to my wife and our children: Sister Michèle; my aunts Geneviève, Maryvonne, Marie-Luce and Françoise; Philippe; Carol; François; Alexandre; Pierre-Emmanuel and Laurence; Jean-Luc and Cathy; Cécile; Jean-Philippe; Philippe; Alain and Darcy; Laurent and many others.

Thank you also to all Ikarian clients who had faith in me and helped me launch a new business that I am enthusiastic about running and fully committed to.

And finally, a huge thank you to my fellow inmates who, through their support, generosity and profound human qualities, enabled me to spend those twenty-five months in detention in the best possible conditions: Georges, Niko, Greg, Jimmy, 'Herbie', Renato, 'Muay Thaï', Filippo, Sanchez, Vladimir, Andrejz, Sasha, 'Fifa', Sam, Tim, Kay and many others. I will never forget you.